Third Edition

Introductory Biological Statistics

Raymond E. Hampton
late of Central Michigan University

John E. Havel
Missouri State University

WAVELAND
PRESS, INC.
Long Grove, Illinois

For information about this book, contact:
Waveland Press, Inc.
4180 IL Route 83, Suite 101
Long Grove, IL 60047-9580
(847) 634-0081
info@waveland.com
www.waveland.com

This book is dedicated to my father, the late Jerome Havel,
who instilled in me an appreciation for numbers,
and my PhD mentor, the late Stanley Dodson,
who showed me how scholarship is best done.
Finally, I recognize the late Ray Hampton,
who wrote the first edition of this book and
entertained us with countless stories and good humor.

Contents

Preface to Third Edition

The purpose of this book is to introduce upper-level undergraduates and beginning graduate students in biology to methods of statistical analysis. As in the previous editions, we use a pragmatic approach, illustrating the many uses of statistics in biology. We provide numerous examples and exercises—some made up, some borrowed from colleagues or from our own work, and others modified from other sources. Understanding the logic behind statistical procedures is very important. We thus introduce students to the theory, state explicitly the assumptions underlying each method, and show how these assumptions can be checked. We illustrate the calculation steps for most procedures and show computer output for many of the same examples.

The third edition has been extensively revised to improve on areas in previous editions where students had difficulties, to correct errors, and update to new standards of technology. The chapter on probability has been thoroughly reworked to make clear the different rules for counting and calculating probabilities prior to introducing the Binomial and Poisson distributions. I have made every attempt to retain the clarity and informality of Ray Hampton's first edition. Toward that end, chi-square tests are moved earlier in the book, to follow chapters on discrete probability distributions and a new chapter on hypothesis testing. Most biology students have been previously exposed to these procedures in other courses, and so this topic makes an effective entry into hypothesis testing. The chapter on the normal distribution has now been expanded to incorporate sampling distributions (e.g., standard error of the mean) and using the normal approximation to make inferences about proportions. The chapter on the one sample t test now includes confidence intervals about the mean, as well as hypothesis testing about a single mean. Paired comparisons immediately follow to show the logical connection to one-sample tests. Chapters 12–14 have changes to topic sequence to improve clarity and balance in chapter length. The chapter on experimental design is now expanded to more clearly illustrate the most important principles and a reprinted article about the famous polio vaccine trial serves as illustration. While retaining the focus on traditional statistical methods, I have also introduced some modern techniques in boxed examples written by experts: computer-intensive resampling methods, multivariate analysis, and meta-analysis. A glossary of statistical terms has been added to the appendix.

Many exercises were replaced to omit excess redundancy and improve student understanding. Different types of exercises now appear in pairs so that the odd-numbered answers for one should help in solving the other member of the pair. Instructors are then free to use the even-numbered exercises for problem assignments and exams. A complete solution manual is now available for instructors, showing all details of the solutions. More integration between hand-worked and computer-based approaches allows better student understanding of how the computer program gets its answers.

We have now bundled the text with Minitab Statistical Software (Release 14 for students), which is fully sufficient for solving all the exercises in this book. Including the software should remove the frustration of many students wishing to use the software away from campus labs, without paying the price for the most current version. The additional cost is modest, keeping the total price of the book inexpensive.

I appreciate the many favorable comments from former students (now exceeding 1,000) and instructors using the second edition, and thank them for pointing out difficult areas and errors. Several department colleagues served as "sounding boards" for ideas and generously provided examples and reviews of the more-difficult chapters. I particularly thank Brian Greene and John Heywood. I also thank the colleagues who wrote boxed examples on resampling methods (John

Heywood, Missouri State University), multivariate analysis (Kim Medley, University of Colorado), and meta-analysis (Jon Shurin, University of California-San Diego).

My research assistant, Risa Wright, did a wonderful job with revising the exercises, offering an elegant proof for chapter 7, and critiquing the rest of the text. Editor Dakota West capably took the rough text and figures and created an attractive book at low cost.

John Havel
Springfield, Missouri
March 2013

chapter one

Why and How We Use Statistics and Graphic Displays

Science is both an organized body of knowledge and a way of knowing. Any student of the sciences will agree that subjects like biology, chemistry, and physics contain a lot of information. What they have in common is the process for generating new knowledge. Any science progresses by developing broad new generalizations (theories) that explain observable events of the physical universe. The essence of science is skepticism: all ideas are evaluated with observable data and theories are developed only after extensive testing of these ideas. By their nature, hypotheses must be subject to repeatable tests that can falsify them.[1] In time, weak ideas are rejected and correct ideas are validated. Statistics plays its role in helping us to more efficiently sort out the good ideas from the bad. Statistics also allows us to quantify our level of uncertainty about what we know.

1.1 What Is "Statistics" and Why Is It Useful?

Statistics can refer simply to "a collection of numerical data." However, most scientists define **statistics** as "the mathematics of the collection, organization, and interpretation of numerical data and the analysis of population characteristics by inference from sampling" (*American Heritage Dictionary*, 1st ed.). Statistics is the science of uncertainty, and its goals include assigning probabilities to the reliability of estimates, to the reliability of conclusions, and to the likelihood of the outcome of future events. Statistics, at its best, is another way of thinking about much of the functioning of the physical universe.

For biologists, statistics is widely used in most subdisciplines—from molecular biology to ecology. Pick up any professional journal and you are bound to run across statistics. In all of these fields, statistics is used by scientists to organize and describe data and to help evaluate the results from experiments. Physiologists use regression models to explore the response of metabolic rates on temperature. Multivariate procedures (chapter 15) are used with gene sequence data to explore the evolutionary relationships within apes, flowering plants, and influenza viruses. Statistics is an important tool for public health, helping disease detectives to track the causes of diseases in populations. Statistics is also widely used in product safety, through guiding the efficient sampling of food for pathogenic microbes. Understanding statistics is also important for educated citizens. Numerical data and conclusions about their meaning are often presented in the media, and sometimes the message is misleading. A firm understanding of statistics provides the reader with a healthy bit of skepticism, for helping to sort the weak from the strong conclusions.

Statistics has several important uses to biologists and to anyone else who investigates events that are less than certain. One use of statistics is to provide a summary of quantitative data. Graphs and tables are employed in most of the chapters in this book. Descriptive statistics reduce a mass of numbers to something more comprehensible, and are introduced in chapter 4. Another use of statistics is to determine, with a specified degree of uncertainty, if the data mean what we think they do about some question of interest. This process refers to inferential statistics, which provides a widely accepted standard for estimating parameters and testing hypotheses. A key ingredient of the scientific method is this comparison of observable data with an idea about how we think nature works. We will examine the nature of hypothesis testing in some detail in chapter 6 and explore a variety of methods in chapters 9–14.

A final use of statistics is to provide a framework for experimental design, which guides the collection of future data and assures that we can reach a defendable conclusion following a survey or experiment. Good experimental design makes sense for all scientific investigations and is particularly important if a large amount of time and resources are involved. Experimental design principles are introduced throughout the book and discussed further in chapter 15.

Caution

Many of us have spent time, energy, and money conducting experiments without giving much regard to the statistical analysis that will be applied to the collected data. Quite often, at the end, there is no test we can properly apply to our data because of the experimental design that was used (or because of the lack of one). Consequently, an inappropriate test is sometimes employed, either through ignorance or desperation. Biologists are thus well advised to think about statistical methods during the early phases of their study.

With any statistical analysis there are certain assumptions about the nature of the data and about how they were collected. Think of these assumptions as rules for when a particular test may or may not be properly used. If our experimental design and data do not follow these rules, we should not use the test! Usually, but not always, violations of these assumptions invalidate any conclusions based on the test. Thus, while it is possible, computationally, to apply many different tests to the same set of data, the validity of the conclusions based on any given test depends on how well the assumptions of the test were met, not on how well the test supports the conclusion we wish to support.

1.2 Populations and Samples

Statistics are often used to make **inferences** (broader conclusions about populations) based on data collected from only a portion of the population. That portion of the population that we collect and measure (or count) is called a **sample**. In statistics the term **population** refers to all the individuals (or measures) of interest.[2] Notice the difference in how the term "population" is used in this statistical sense and how "population" is usually used by biologists and sociologists. To a biologist, a population represents all the organisms that belong to the same species and live in the same geographic area. To a sociologist, a population refers to some group of human beings (e.g., US population aged 15–30 years old). Notice also how the word "sample" differs in statistics from some other uses in biology. A biologist sometimes refers to a blood sample (substance taken from one individual) or a water sample collected on one day from a lake. In statistics, we are interested in the subset of the population of interest. So, the physiologist has a sample of blood glucose concentrations from 15 randomly-selected patients on a new drug for diabetes, and the limnologist has a sample of phosphate concentrations from 20 different lakes. Some examples of statistical populations and samples are shown below in table 1.1.

Consider the cholesterol example in table 1.1. If we wish to measure the effect of a medication to treat high cholesterol in middle-aged adults, the population of interest is the cholesterol levels of all middle-aged adults (or more accurately, the change in cholesterol concentration before and after taking the medication in this group of adults). One must always take care to specify the limits of the population being investigated. In our example, it's well

---

Table 1.1   Some Examples of Statistical Populations and the Samples Drawn from Them

Question of Interest	Statistical Population	Sample
What is the mean height of females at Overachiever University?	Heights of all 30,000 female students at Overachiever University	Heights of 50 randomly-selected female students from Overachiever University
Is the proportion of the white-eye phenotype in *Drosophila* following a monohybrid cross = ¼?	Proportion of the white-eye phenotype in *Drosophila* following all possible monohybrid crosses	Proportion of 80 flies with the white-eye phenotype from a monohybrid cross
What is the effect of Lipitor on cholesterol concentration in middle-aged adults having high blood cholesterol levels?	Change in cholesterol concentration (before-after) in all middle-aged adults having high blood cholesterol levels	Blood cholesterol concentration before and one month after taking Lipitor in 200 randomly-selected patients having high cholesterol

known that cholesterol levels depend on genetics and diet. So, we may wish to restrict our attention to a particular area, say the United States. It is always important to recognize how well the sample represents the true population of interest.

To consider populations and sampling more fully, consider a simpler example, estimating mean height of female university students.

Example 1.1
A Random Sample from a Population

The heights of a sample of 50 female students in a midwestern university were determined. The sample was selected from among all female students in the university by randomly picking their names from the student directory.

We might consider this measured group to be a sample of a larger population, but what population do they represent? Is the average height of this sample a reasonable estimate of the average height of all women everywhere? The only population about which we can reasonably draw conclusions is the population of female students in this university, because this is the population from which our sample was drawn. If we are willing to assume that these college women are fairly typical in stature of midwestern women in their late teens to early twenties, we could generalize our conclusion to include this larger group. Note that the key phrase here is "willing to assume." Our generalization to a larger population than the one that was sampled would be accurate only if our assumption is correct! In a purely statistical sense, we have no basis for concluding anything about the height of women who are not members of the sampled population.

1.2.1 Random and Biased Samples

Notice the term "random" in example 1.1. Randomness does not imply casual, haphazard, or unplanned. Rather, **randomness** means that each possible sample of the same size that could conceivably be drawn from this population has an equal probability of being drawn. When a sample is not randomly selected, the sample is **biased**. Let's say that we had collected our height measurements at a station near the gymnasium; we might be more likely to encounter athletes, who may be taller on average. Clearly, this would be a biased sample.

We are all aware of the word *bias* used in a social sense, such as racial bias in renting apartments to various minority groups. Racial bias generally means

that individuals of these minority groups are underrepresented among the tenants of the apartments. This bias is often assumed to be due to racial discrimination, and, by today's standards, such racial discrimination is both unethical and illegal. (Interesting experiments can be devised to confirm whether this discrimination assumption is true or not!)

Consider another example of bias that we can examine in more depth.

Example 1.2
A Biased Sample from a Population

A large university wanted to illustrate the success of its graduates by publicizing the average annual income of its former students. They selected alumni 10 years after graduation and attempted to contact each individual by telephone. Although able to contact only 20% of this group, the university averaged the results and reported that their alumni earned over twice the average for this age group in the general population. The university thus assumed or implied that this sample of 20% was a random sample from the entire population.

Telephone surveys are widely used in the social sciences, and, at first glance, seems a good way to obtain the alumni income data. In this case, the university sampled 20% of this alumni group for their incomes. However, this was not a random sample, as the 80% missing from the sample included alumni who could not be reached (or would not answer the question). Those responding to the survey were likely a biased sample from the population of interest. We could imagine that people who were easier to reach by phone would include a higher fraction of professionals and those listed in who's who directories. The survey would be less likely to reach those who were homeless or substance abusers, groups which are less likely to earn the larger incomes.

Biased sampling is also a problem in biology. For example, snake population size is sometimes estimated by conducting road surveys. The number of snakes spotted crossing the road should be proportional to population size and one might think that sex ratio could be determined this way as well. However, biologists must be careful in how they interpret such data, since this survey technique is dependent on the activity levels of the snakes. For instance, in the spring nearly all snakes observed crossing the road are adult males. We could be tempted to conclude that this population has very few females, when in fact only the adult males are actively searching for mates! The females are out there, but are not fairly represented by this survey technique in the spring.

An important rule for the interpretation of statistics is to be aware of the population of interest versus the population actually sampled. When these populations fail to agree, we either need to change our sampling strategy or modify our conclusions about the results. For the snake example above, making this distinction about sampling requires knowledge about the natural history of the animal.

1.2.2. Collecting Random Samples

How does one obtain a random sample? There are a number of ways to do this. Most of us have drawn names or numbers from a hat, or have seen a commercial for a lottery where one numbered ball is selected from a drum. The process is designed so that each number or name has the same chance of being drawn.

Another method, which helps illustrate the process, is the use of a random number table (see appendix A, table A.10). In example 1.1 we could have assigned a number to each woman in the university and then consulted the random number table to decide which individuals to include in the sample. To do so, we close our eyes and touch the table to select a starting number. We then collect consecutive numbers from left to right. The fact that each group of numbers in this table consists of 5 digits is of no significance. This is done simply to make the table a bit easier to read. If we wish to select random numbers of three digits each, we simply read them in groups of three.

Finally, we can select random numbers by using a computer to generate random samples. *Minitab Statistical Software* is bundled with this textbook, so let's consider how to generate random samples with this program (box 1.1).

1.3 Graphical Representation of Numerical Data

Organizing numerical data into logical and clear tables and graphs is an essential first step in any data analysis. Besides preparing for further analysis with hypothesis tests, such displays help researchers to check assumptions about their data (e.g., normality), view trends worth exploring in more detail later, and catch mathematical or data entry errors. Thus, a general rule of thumb is to *always plot your data*.

A wide variety of graphics can be used to display numerical data, and the particular choice of

---

Box 1.1

With *Minitab*, the *sample* command allows us to select a random sample of a particular size from a larger dataset. For example, suppose we have 400 study plots and wish to collect plants from 20 randomly chosen plots. We could designate each plot with a number placed in column C1. (Hint: using the *set* command can do a consecutive series quite easily.) To choose our random set of 20 study plots and place their numbers in column C2, simply say: <sample 20 c1 c2>. (Do not type the brackets.) Alternatively, using menus, choose "calc," "random data," then "sample from columns." This particular sample was done without replacement, appropriate for our goal here. If we wished instead to sample with replacement, we would have to specify that with the "replace" subcommand.

These randomization procedures can now be done so quickly that thousands of computer-drawn samples can be collected from an original collection of data, and the distribution of those samples used to test hypotheses about the original population. This topic is described in chapter 15 (Guest Box 15.1, pp 158–159).

---

graph depends somewhat on the type of data being represented. You have undoubtedly seen or used many of these graphs before. Here we'll briefly view some common graphs; in later chapters you will see similar graphs again.

Pie charts[3] (subdivided circles) are commonly used to display data that have been converted to %. For instance, you might display a household budget as the % of monthly income that goes to rent, gas, food, etc. Such graphs are commonly used in the media. Biologists use pie charts to display such things as time spent by animals in different activities and proportion of diet from different foods (fig. 1.1). You can see that these hedgehogs eat mostly beetles. Notice also that the size of the pies indicate that hedgehog B ate more total food than hedgehog A.

Bar graphs show different bars on the x axis, with each bar's height shown on the y axis. Bar graphs are commonly used to display frequency distributions. Frequencies are just a tally of something that has been measured. For example, in ecology we might establish a grid of uniform plots and count the number of plants or animals per plot. Figure 1.2 shows that most plots had 0 flatworms, but up to 3 flatworms were found in a few of the plots. Frequency distributions will be explored in detail in chapter 3.

Bar graphs can also illustrate a measurement variable (on the y axis) against categories on the x axis. We can compare the differences between groups by showing their averages with the height of the bars, and lines illustrating amounts of varia-

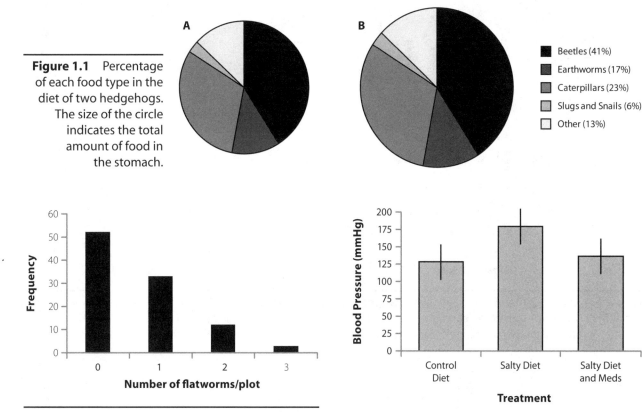

Figure 1.1 Percentage of each food type in the diet of two hedgehogs. The size of the circle indicates the total amount of food in the stomach.

Beetles (41%)
Earthworms (17%)
Caterpillars (23%)
Slugs and Snails (6%)
Other (13%)

Figure 1.2 The abundance of terrestrial flatworms in 100 surveyed plots.

Figure 1.3 The effect of diet and medication on systolic blood pressure. Data are represented as the mean ± SE. (Imaginary data)

tion within each group. In the imaginary data shown in figure 1.3, people on the salty diet alone had substantially higher blood pressure (mean about 180) than those in the control group and those with both salty diet and medication (mean about 130). Notice that the latter two groups appear no different. Such illustrations are handy displays to support results from formal hypothesis tests, such as *t* tests (chapter 10) or Analysis of Variance (chapter 11).

A **scatterplot** displays the relationship between two measurement variables. Examples are shown in figure 1.4. Graph A shows that the two different types of scores appear to be correlated with each other. We wouldn't want to say that one causes the other; more likely, students with high scores on one exam have the family environment and study skills that contribute to high scores on the other exam. Graph B shows the dependent relationship of metabolic rate (the response variable) on environmental

temperature in experimental chambers. The strong dependence of metabolic rate on temperature implies a causal relationship. Notice that the regression line is shown on this graph, but not on graph A, where there is no causal relationship implied. Similar graphs are displayed in chapters 13 and 14, where they provide supporting evidence with formal statistical tests (correlation and regression analysis).

Finally, the **time-series graph** illustrates patterns over time. The graph shown in figure 1.5 shows the change in daily air temperatures over the course of the summer. Notice that the lines (mean and maximum) each connect the individual daily measures (individual dots for each day not shown). Connecting the daily measures is typical for time-series graphs. Notice the difference between this graph and figure 1.4B, where the individual points are not connected, but the regression line is shown.

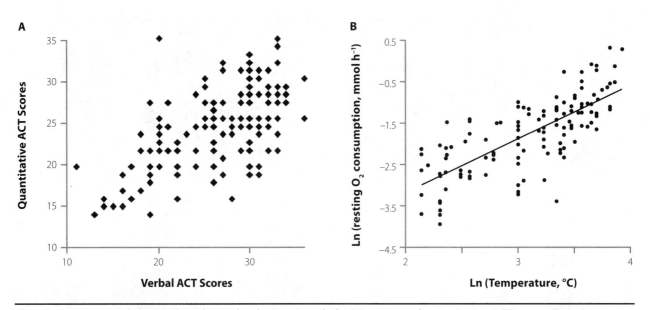

Figure 1.4 Scatterplots. (A) The relationship between verbal ACT scores and quantitative ACT scores (imaginary data) (B) Log/log plot with fitted least squared regression line demonstrating the relationship between resting metabolic rate and temperature in teleost fish (Data from Clarke and Johnston. 1999. *Journal of Animal Ecology* 68: 897)

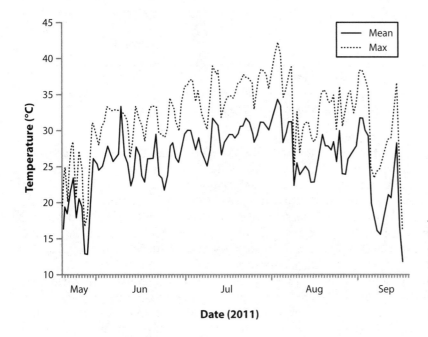

Figure 1.5 Example of a time-series graph. Air temperatures from 2011 for Springfield, Missouri (daily average and daily maximum). (Data from the National Weather Service, graph prepared by K. Reynolds)

Key Terms

bar graph (bar chart)
biased sample
pie chart
population (statistical population)
random sample
sample
scatterplot
statistics
time-series graph

Exercises

1.1 Select five simple random samples of 10 bluegill sunfish lengths from the data for 888 fish in Digital Appendix table 1. (If you are using a computer—the far easier method—store these samples in separate columns; if not, write down the lengths for fish having an ID number equal to each random number.) Repeat this process for samples of 20 fish and then 30 fish. We will explore these random samples further in later chapters, so save the data in a spreadsheet you can find when you need it.

1.2 Using the procedure in exercise 1.1, select five simple random samples of 10, 20, and 30 male mosquito fish lengths from the data in Digital Appendix table 2. Collect similar random samples of female mosquito fish lengths. Notice that the data show a code for sex.

Notes

[1] For a fascinating description of the nature of science, see the article "Science and the Courts" (Ayala, F. J. and B. Black. 1993. *American Scientist* 81:230–239.)

[2] More precisely, a **population** refers to all objects of a particular kind in the universe or in some designated subdivision of the universe.

[3] Note that many students and teachers (and the *Excel* program) use the term "charts" to refer to "graphs." For scientific publications, the term "graph" is the general term (exception pie chart), and both are subsumed under the more general term "figure." Figures can include graphs (or charts), schematic drawings, and photos—any published information except tables and text (and computer databases).

chapter two

Data Measurement and Management of Numbers

Before we get too far with statistics, it's important that we develop a common vocabulary about the sorts of things we measure and count, as well as about how the numbers that are generated can be organized. In this chapter we will examine the different sorts of variables encountered in statistics and how they are represented. We will also briefly see how data can be managed in an efficient manner. In chapters 3 and 4 we will examine different ways numbers are summarized and presented through graphics and descriptive statistics. A good foundation of data description makes hypothesis testing much more understandable, both in this course, and as a general practice.

2.1 Variables

Variables are characteristics that may differ from one member of a population to the next. Data are the values of these variables for individual members of the population.

2.1.1 Types of Variables

Variables can be grouped according to several characteristics, such as whether they represent measurements, ranks, or simple categories. Measurements may be either discrete or continuous. Consider a couple of examples.

In example 2.1 tree height is a variable. A variable such as this is a **continuous measurement variable** because it may assume any imaginable value within a certain range. The number of decimal places reported for such a measurement depends on the measuring instrument (e.g., weight on a bathroom scale versus analytical balance). Measurements of

---

Example 2.1
A Continuous Measurement Variable

The heights of 6 pine trees in a certain area of forest were measured, with these results:

Tree Number	1	2	3	4	5	6
Height (m)	31.3	29.1	32.6	19.5	37.8	29.1

---

length, mass, time, temperature, concentration, and so on, are examples of continuous variables.

In other situations, we may collect data on phenomena that occur in only discrete integer steps, where intermediate values are not possible, as in example 2.2.

Example 2.2
A Discrete Measurement Variable

In investigating the litter size of garter snakes, the following data were collected:

Snake Number	1	2	3	4	5
Litter Size	5	8	6	9	7

Since fractions of baby snakes do not exist, we would never expect to obtain a measurement that is not a whole number. Variables of this kind are called **discrete measurement variables**.

Still other variables cannot be measured on a scale in which the intervals or units have a consistent relationship to each other, yet the observations made on such variables can be ranked with respect to their relative magnitude. These are called **ranked variables** (sometimes called **ordinal**). The position of an individual in a dominance hierarchy, in which individuals may be ranked with respect to their relative position in the group, is a good example of a ranked variable. Suppose that a wildlife biologist is studying the dominance hierarchy in male lions. (Their dominance is important for such things as controlling territories and access to mates.) The biologist records contests she observes between pairs of males and scores who wins each contest. Based on these data, she places the males in this order, from most dominant to most submissive: Yellowhead > Spooky > Bigfoot > Shyone. For ranked variables, the actual distance between any two adjacent categories (in this case, names of lions) is unknown and need not be the same between all adjacent pairs. For instance, Yellowhead may be just barely dominant over Spooky, while Spooky is strongly dominant over Bigfoot. Customarily, such measurements are "scored" using numerical symbols. The "smallest" or lowest ranking item ("Shyone," in this example) receives a score of 1, the next smallest ("Bigfoot") a score of 2, and so on. "Yellowhead," in our example, would receive a score of 4. Ranked variables are always measured on an ordinal scale.

Variables that can neither be expressed quantitatively nor ranked with respect to relative magnitude are called **attributes** (also called nominal). Male or female, red or white, alive or dead, and so on, are examples of attributes. One may designate the class of objects to which an individual belongs, but values such as "greater than" or "less than" have no validity when applied to such attributes. In statistics, we commonly encounter frequency data, where number of individuals are classified into different categories of such attributes.

In summary, we have identified basic variable types. Examples are shown below (table 2.1). What are some other examples?

Table 2.1   Some Examples of Variable Types

Variable type	Examples
Attribute (Nominal)	Fruit fly phenotype: white eye or wild type Sex of snake: male or female
Ranks (Ordinal)	Status of patient: weaker, stable, better Pollination sequence: first, second, third
Discrete Measurement	Number of points on deer antlers Number of fleas on a dog
Continuous Measurement	Body temperature (°C) Weight of a warthog (kg)

Occasionally, measurement data may be **transformed**. Such a variable is the result of performing some mathematical operation on the original variable or of having a measuring device perform such an operation. An example of such a device is a pH meter, which displays the negative logarithm of the hydrogen-ion concentration in a solution (pH = −log[H^+]). The actual variable being measured is the hydrogen-ion concentration. The transformed variable is pH.

Example 2.3
Transformed Measurements

The pH of a solution is a transformed expression of hydrogen ion concentration.

[H^+]	10^{-5}	10^{-6}	10^{-7}	10^{-8}	10^{-9}
$\log_{10}$ [H^+]	−5	−6	−7	−8	−9
pH	5	6	7	8	9

Mathematical transformations are frequently used during advanced statistical analyses (e.g., see chapter 11: ANOVA, when assumptions are not met). Rescaling a graph (such as placing the original values on log scale) is equivalent to doing a log transformation on the original data.

Other times a reported variable may be computed from two continuous measurements. Many rates are such **derived variables**. If we measure how far a snail travels, we have measured a variable called distance. If we measure how long it takes to get there, we have measured another variable called time. If we combine these two variables and measure how far the snail travels divided by the time of its trip, we have now measured a derived variable called velocity. Examples of derived variables are ratios, percentages, and rates.

Often, variables are described in terms of how we think they interact with other variables (the stuff of hypothesis testing!). **Response variables** are those whose variation we suspect depends on some other variable, called a **predictor variable**. For instance, in regression analysis (chapter 14) we may be interested in testing the influence of drug dosage on blood pressure. Here, blood pressure is the response variable and drug dosage the predictor. You may have also referred to these as dependent and independent variables, respectively. Both sets of terms are correct.

All these descriptions of variables are part of the common vocabulary in statistics. So you should know these terms and expect to see them frequently throughout this book.

2.2 Data

Data are individual measurements or observations of a variable made on individual units of the population under study. Data are also sometimes called variants or observations. In the pine tree example, each tree is a unit of the population; height is the variable; and the height of an individual tree is a datum, variant, or observation.

2.2.1 Representing Data with Symbols

A variable is conventionally designated with a letter, typically x, and individual data (observations, variants) as $x_1, x_2, x_3, \ldots, x_n$, where x_1 is the first datum, x_2 the second, and x_n the nth observation in the series. When two variables are under study, one is usually designated as x and the other as y. But this convention is not sacrosanct! Professionals working with numbers might represent variables by other letters; however, we must stay consistent once a variable has been defined. The total number of observations of a variable in any given sample, the **sample size**, is designated as **n**.

2.2.2 Significant Figures and Rounding Rules

Before we go further, let's consider a common practical problem with reporting the values of continuous measurements. Occasionally, following some calculations we are left with a large number of digits. For instance, suppose we are interested in the respiratory physiology of elephants and have made the following measurements. Big Burt, who tips the scales at 1043 kg, consumes 202 liters of oxygen per minute (202 L O_2 min.$^{-1}$). According to my calculator, his weight-specific respiratory rate is 0.1936721 L O_2 min.$^{-1}$ kg^{-1}. Despite what my calculator reports, the weight-specific respiratory rate should be reported only to three significant figures. Whatever measurement has the fewest significant figures sets the limit to the number of significant figures in the derived variable. In the case of Burt, his weight-specific respiratory rate should be reported as 0.194 L O_2 min.$^{-1}$ kg^{-1}. Notice that the leading zero doesn't count as a significant figure, since this number could just have well been written in scientific notation as 1.94×10^{-1}.

Rounding numbers follows some simple and important rules, listed below in table 2.2.

To repeat a key rule, when doing a series of calculations, *save all the numbers in memory* (either in the calculator or computer) until the final answer is complete. It is then common to first show the answer with more than enough digits, and then round to the proper number of significant figures. Later, as you are doing practice exercises, you will notice that the answer key often shows more than the correct number of significant figures. This extra information is to make it easier for you to know that your answers are on the right track. In a formal report, you should round the final answers.

Table 2.2   Rounding Rules

Intermediate results	Never round off intermediate results. This practice introduces rounding errors. Use your calculator memories to *store all* intermediate values. Round only at the end.
Precision of measurement	The number of digits displayed for a continuous measurement indicates the precision with which it was measured and hence its significant figures. (However, leading zeros do not count.)
Significant figures in a final answer	The precision of the final answer from a series of calculations represents the precision of the original measurements. Typically, an average will have one more significant figure, if the sample size was 10 or more.
Rounding	Round up (add one to the last significant figure) when the remaining numbers are anything greater than 5 (e.g., 51); round down (keep the last significant figure the same) when the remaining numbers are anything less than 5 (e.g., 49); when the remaining number is exactly 5 (e.g., 50), round up when the previous number is odd and round down when the previous number is even. **Examples** (round to 3 significant figures): 1.852 → 1.85 1.856 → 1.86 0.1852 → 0.185 1.855 → 1.86 1.865 →1.86
Calculations from exact values are not rounded	Calculations of probabilities from exact numbers should not be rounded. For example, in the roll of a die (with six faces) the chance of any one outcome is 1/6 or 0.166666 . . . , best written as 1/6 (or express at least six significant digits).

2.2.3 Converting Data from Measurement to Ranks

Sometimes, we may wish to convert measurements into ranks. For instance, creating a sorted list allows determining descriptive statistics such as medians and quartiles (chapter 4). Similar data conversions are done when using nonparametric statistics (e.g., chapter 10, Mann-Whitney U test). Tree height is a continuous variable. Suppose we wanted to convert these measurements to ranks. To illustrate ranking, consider example 2.1, which is reproduced here:

Tree Number	1	2	3	4	5	6
Height (m)	31.3	29.1	32.6	19.5	37.8	29.1

The shortest tree in the group is tree 4, so we assign it a rank of 1. The next shortest are trees 2 and 6 (both with a measurement of 29.1); they are tied for ranks 2 and 3, and hence given the average of these two ranks (2.5). The remaining trees are ranked until we have ranked each tree with respect to all the others. Tree 5, being the tallest, is given a rank of 6. Our complete data now look like this:

Tree Number	1	2	3	4	5	6
Height (m)	31.3	29.1	32.6	19.5	37.8	29.1
Height rank	4	2.5	5	1	6	2.5

Note that when we do this sort of conversion, we lose some information originally contained in our data. With ranks alone, we now know only which tree is taller or shorter than which others. We do not know how much shorter or taller, nor do we know the height of any individual tree. But sometimes that is all we need to know.

2.3 Data Management and Statistical Software

While studying statistics, you will have access to calculators, spreadsheets, and software packages dedicated to statistics. These modern inventions are a step up from the abacus and slide rule, but they don't think very well! That's your job. Setting up the problem on paper or in your head is always the critical first step. Remember that computers are very efficient at giving wrong answers, as well as right answers. A very important skill for students (and their instructors) is being able to quickly spot errors. One way to do that is to make a "ball-park estimate" ahead of time and then see whether or not later answers make sense. A simple graph is very helpful in this regard. In chapter 3, we will see how to construct a histogram, which allows us to visualize the average and variation in a set of data.

When working with numbers, one has a choice among many different tools. Sometimes hand calculations are simplest, and these may be the only method available, such as while thinking in a restaurant or trying to impress your boss! A cheap "scientific" calculator is useful for doing operations on small masses of data and will be indispensable for working on problems in this book. Besides the usual arithmetic operations, knowing how to use the storage memories and parentheses can save you much time and will reduce the chance of rounding errors. Although these calculators also have some statistical functions, computers are much better at this process. Spreadsheets (e.g., *Excel*) are very useful for organizing numbers, preparing tables, and doing simple repeated operations. These spreadsheet packages also do graphing and inferential statistics, but not very well. Numerous computer packages dedicated to statistical analysis are now available (e.g., SAS, SPSS). *Minitab Statistical Software* is versatile and easy to use. We have chosen to bundle this textbook with *Minitab*, and have also shown some examples in various places in the book. Such statistical packages do data manipulations, descriptive statistics, and a wide variety of inferential tests. Statistical packages also do graphing, sufficient for the uses of this book. Most professionals move between different programs. For instance, within a *Windows* landscape, one might manage numbers and create tables in *Excel*, copy and paste numbers into *Minitab* to run statistical analyses, and then copy and paste tables and statistical output into *Word*, to support writing a report. (By the way, we did these things repeatedly in preparing this book!) Copying and pasting numbers into a dedicated graphics package, such as *Sigmaplot*, allows preparation of publication-quality graphics.

Key Terms

attribute (nominal)
continuous measurement variable
data (plural)
datum (singular)
derived variable
discrete measurement variable
predictor variable
ranked variable (ordinal)
response variable
sample size (n)
significant figure
transformation
variable
variant

Exercises

2.1 Following are the heights (in centimeters) of a small sample of male humans:

187 171 181 180 178 171 174 177 172 178 182

  a Is this a discrete or continuous variable?

  b The calculator gives an output for the average height as 177.363636363… Round the average to the correct number of significant figures. *177.4*

2.2 Following are the lengths (in millimeters) of a small sample of largemouth bass:

210 325 285 402 350 240 409 330 295 325 256

  a Is this a continuous or discrete variable?

  b The calculator gives an output for the average height as 311.545454545… Round the average to the correct number of significant figures.

2.3 An ecology class determined the number of ant lion pits in a sample of 100 randomly selected one-meter-square quadrats. The results were as follows. Is this variable discrete or continuous?

Pits/Quadrat (X)	Number of Quadrats Containing X Pits (Frequency)
0	5 *= 0*
1	15 *= 15*
2	23 *= 46*
3	21 *= 63*
4	17 *= 68*
5	11 *= 55*
6	5 *= 30*
7	2 *= 14*
8	1 *= 8*

Discrete

299 = 2.99
100

2.4 Garter snakes respond to an overhead moving object by exhibiting an "escape" response. The intensity of this response may be measured on a somewhat subjective scale ranging from 3 to 0. A rapid movement of the snake from one location to another is given a score of 3, a movement involving at least one-third of the snake's body but not resulting in relocation of the animal is given a score of 2, a slight movement of the head only is given a score of 1, and no visible response is given a score of 0. Fifteen snakes were tested for this response and received the following scores. What variable type are these data?

3	2	3
1	3	2
3	3	1
0	2	3
2	0	2

2.5 Convert the following male heights (given in centimeters) into ranks.

Male ID	1	2	3	4	5	6
Height	181	202	190	185	190	200
Rank	*1*	*6*	*3.5*	*2*	*3.5*	*5*

3+4=3.5
2

2.6 Convert the following number of bullfrogs per pond into ranks.

Pond ID	1	2	3	4	5	6	7	8	9	10	11
Bullfrogs	34	65	23	34	18	20	15	70	15	18	34
Rank	—	—	—	—	—	—	—	—	—	—	—

2.7 For the following situations, identify the type of variable (discrete, continuous, rank, attribute). Also, indicate if any of the measurement variables are transformed or derived.

  a The velocity of an enzyme-catalyzed reaction of substrate in micromoles converted per milligram of protein per minute. *continuous, derived*

  b The number of male bullfrogs in a pond. *discrete*

  c Sex of a salamander: male, female, intersex. *attribute*

  d pH and pK$_a$. *transformed, continuous*

  e The velocity of snails, in furlongs per fortnight. *continuous*

  f The order in which individuals of a wolf pack feed on a kill. *Rank/ordinal, derived*

nominal

2.8 Using conventional rounding rules, round the following numbers to three significant figures. (Hint: it is sometimes helpful to express in scientific notation.)

  a 106.55

  b 0.06819

  c 3.0495

  d 7815.01

  e 2.9149

  f 20.1500

  g 20.2500

Frequency Distributions

Large quantities of data are generally difficult to interpret by themselves. In this chapter and in the next on descriptive statistics we consider some basic techniques for organizing and presenting data to make them more readily understandable. Graphic techniques are widely used for data exploration by investigators and also to display summary data in published literature. We focus on frequency distributions in this chapter and then briefly introduce some other graph types that will appear in later chapters.

3.1 Frequency Distributions of Discrete Variables

Frequency distributions display a summary of the number of observations which take on specific values or range of values. The **frequency** is a count of the number of observations in a category. Frequency distributions may be shown either as a table or a graph, and the data summarized can be from either discrete or continuous measurement variables.

You will recall that discrete variables may assume only integer values and that intermediate values are not possible. Example 3.1 presents raw data for a discrete variable and then shows how these data may be organized.

Clearly data in this form are rather difficult to interpret. From glancing through the numbers it would seem that the quadrats contained about 2 or 3 plants each, except that many contained no plants at all. As an initial step in analyzing data, it is frequently advantageous to group the data into a frequency distribution. In this case we would determine how many quadrats contained 0 plants, how many contained 1

Example 3.1
Frequency Distribution of a Discrete Variable

The raw data in table 3.1 show the number of maple seedlings that were present in 100 one-meter-square quadrats. (A quadrat is a square sampling unit used by ecologists to measure the abundance and distribution of plants and stationary animals.)

Table 3.1 Maple Seedlings per One-Meter-Square Quadrat

0	1	0	2	0	0	1	2	1	1
0	1	2	5	1	0	2	1	1	2
1	2	3	4	3	4	2	1	0	5
0	0	2	0	1	0	0	0	2	0
1	3	4	2	1	0	0	1	0	1
2	3	5	0	0	0	1	2	3	4
3	3	0	2	0	4	5	0	0	0
0	1	0	0	1	0	1	0	0	1
0	1	0	1	3	1	5	1	1	3
4	3	2	0	0	4	1	2	1	1

Data from M. Hamas

---

plant, how many contained 2 plants, and so on. We would then tabulate the results in the manner shown in table 3.2 on the following page. This arrangement makes the data much easier to deal with. We can see at a glance that 35 quadrats contained 0 seedlings, 28 contained only 1, and so on.

A frequency distribution is often represented in the form of a graph, which makes the information it contains even more understandable. Graphs of frequency distributions for discrete variables are called **bar graphs.** Figure 3.1 is such a bar graph, constructed

Table 3.2   Maple Seedlings per Quadrat

Number of Plants/Quadrat	Frequency
0	35
1	28
2	15
3	10
4	7
5	5

from the frequency distribution in table 3.2. Here, we are plotting a measurement variable along the x axis (in this case, the number of plants per quadrat) against the number of times that variable occurred (its frequency) on the y axis. Note that the bars in figure 3.1 representing the various frequencies do not touch each other. This is because a discrete variable is involved, and intermediate values are not possible. Discrete variables are conventionally plotted this way.

3.2 Frequency Distributions of Continuous Variables

You will recall that continuous variables may assume any imaginable value between certain limits. Accordingly, their graphic and tabular presentations are somewhat different from those of discrete variables. We will need to pool together values which fall within specific class limits (called classes or "**bins**").

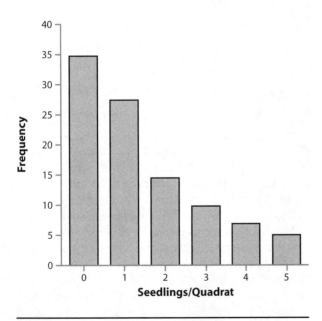

Figure 3.1   Maple seedlings per 1 square meter quadrat (an example of a bar graph). (Data from table 3.1)

Example 3.2
Frequency Distribution of a Continuous Variable

A simple random sample of 172 adult male mosquito fish was collected from a large population, and the total length of each fish was measured to the nearest millimeter. These data are shown in table 3.3.

Table 3.3   Total Lengths (in mm) of 172 Male Mosquito Fish

30	22	23	28	19	22	21	23	22	24
17	22	24	20	20	23	20	18	21	25
19	22	21	20	20	23	22	18	22	21
22	24	17	22	23	21	21	19	24	20
19	22	22	21	23	20	21	20	20	23
22	20	20	24	24	24	18	19	21	21
24	19	22	23	24	28	22	24	21	20
23	23	22	22	21	21	22	20	21	24
23	21	25	25	24	25	23	23	22	20
19	22	23	22	24	24	19	20	24	24
22	23	25	22	20	20	24	20	23	22
22	21	20	21	24	20	24	21	21	22
19	22	23	22	22	22	20	23	22	22
22	21	23	23	20	21	21	20	21	22
19	22	20	20	21	19	24	22	21	20
20	22	22	23	20	19	24	22	21	21
23	29	21	21	21	21	24	23	24	22
24	25								

Once again, inspection of the data in this form makes interpretation difficult. Length is a continuous variable, and the information contained in a frequency distribution of such a variable is somewhat different from that contained in a frequency distribution of a discrete variable. A continuous variable may assume any value within a certain range, while a discrete variable may not. Thus, when constructing a frequency distribution of a continuous variable, it is customary to group measurements into classes (bins) containing only individuals that fall within a certain range of the variable under consideration. For example, we might group all of the male mosquito fish lengths in table 3.3 that measured 17 mm and 18 mm into one class, those that measured 19 mm and 20 mm into another, and so on. Keep in mind that measurement was to the nearest millimeter, so that, in effect, 17 mm really includes individuals from 16.5 mm to 17.5 mm, and so on. Thus, our 17 mm to 18 mm class actually includes individuals from 16.5 mm to 18.5 mm (a range of 2 mm) and our 19 mm to 20 mm class includes individuals from 18.5 mm to 20.5 mm (also a range of 2 mm). In this case we are using a class interval of 2 mm. The choice of class interval is somewhat arbitrary and depends on the number of measurements and on the range of their values. Too many classes tend to be confusing, while too few convey too little information. Trial and error is not a bad way of arriving at a satisfactory class interval! Once a class interval is chosen, a value halfway between the smaller and the larger limit of each class (the midpoint) is designated as the **class mark**. For example, in the 16.5 mm to 18.5 mm class above, 17.5 mm is the class mark. The class mark for the 18.5 mm to 20.5 mm class interval is 19.5 mm, and so on.

Table 3.4 is a frequency distribution of the data in table 3.3, using a class interval of 2 mm. The smallest fish in the sample was 17 mm (actually 16.5 mm to 17.5 mm), which tells us that the first class includes all individuals that were between 16.5 mm and 18.5 mm. In other words, it includes all of those individuals that had a measured length of 17 mm or 18 mm. The second class includes individuals between 18.5 mm and 20.5 mm, and so on. Notice that the sum of the frequencies is 172, the number of fish that were measured (table 3.4).

We implied earlier that doing a frequency distribution by hand is a bit of an art. However, there are some guidelines we can follow. The class interval width is simply the range of the data divided by the number of classes (also called "bins"). The number of bins depends on the sample size. For example, with 100 measurements, we can usually manage about 10 bins. So, if our data ranged in value from 50 to 350 (a range of 300), the interval width would be $300 \div 10 = 30$. We would then make the first bin 49.5–79.5, the second bin 79.5–109.5, and so on.

3.3 Histograms and Their Interpretation

Consider again the frequency distribution displayed in table 3.4. This table is useful for organiz-ing data and doing calculations, but is not so easy to visualize. Here a graphic is helpful. A graph that illustrates a frequency distribution for a continuous variable is called a **histogram**. Figure 3.2 on the following page shows two histograms of these data. Notice that a histogram has no space between adjacent classes. This reflects the basic nature of a continuous variable. In contrast, the bar graph shown earlier (fig. 3.1) has spaces between the classes, consistent with the nature of the discrete variable.

Notice also the difference between the two graphs plotted in figure 3.2. In what ways are they different? To your eye, which graph appears more informative?

Frequency distributions plotted as histograms are very useful for data inspection. Often, we would like to know if the distribution is symmetric about its peak (**mode**) or if the distribution has a long tail in either the left side or right side of the peak (a **skewed distribution**). An example of a skewed distribution is shown in figure 3.3 on the next page. Notice the long tail to the right.

Another characteristic we can observe in a frequency distribution is whether or not the distribution contains one peak or more than one peak. For example, examine the histogram for a large sample of lengths from both male and female mosquito fish (fig. 3.4 on the following page) and notice that this distribution appears to be **bimodal**.

This is an interesting result. To explore further, we can individually plot males and females (fig. 3.5 on p. 19). Notice the histograms indicate a single mode for each of the genders. Notice also that the female distribution is slightly skewed, indicating a smaller number of very large individuals. Skewed data are quite common in many areas of science. Data inspections such as these become very important when we check assumptions, such as normality (chapter 6).

The histogram also allows us to visualize the value of the mean (arithmetic average), as well as the amount of variation (spread) around the mean. Compare the male and female distributions in figure 3.5. Which gender tends to be smaller? Which

Table 3.4   Total Length of a Sample of 172 Male Mosquito Fish (in mm)

Measured Length	Implied Length	Class Mark (x)	Frequency (f)	f*x	Cumulative Frequency
17–18	16.5–18.5	17.5	5	87.5	5
19–20	18.5–20.5	19.5	40	780.0	45
21–22	20.5–22.5	21.5	70	1505.0	115
23–24	22.5–24.5	23.5	47	1104.5	162
25–26	24.5–26.5	25.5	6	153.0	168
27–28	26.5–28.5	27.5	2	55.0	170
29–30	28.5–30.5	29.5	2	59.0	172
		SUM	172	3744.0	

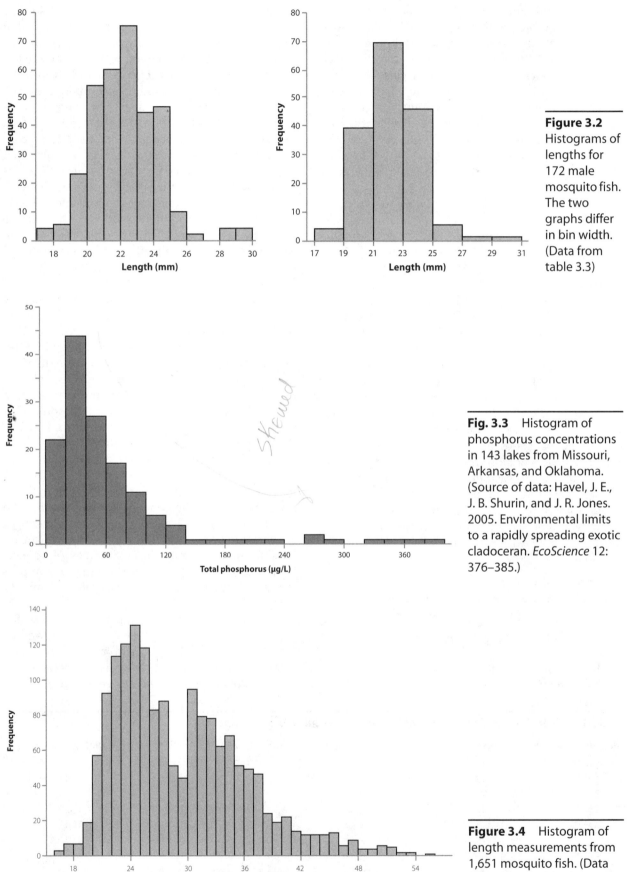

Figure 3.2 Histograms of lengths for 172 male mosquito fish. The two graphs differ in bin width. (Data from table 3.3)

Fig. 3.3 Histogram of phosphorus concentrations in 143 lakes from Missouri, Arkansas, and Oklahoma. (Source of data: Havel, J. E., J. B. Shurin, and J. R. Jones. 2005. Environmental limits to a rapidly spreading exotic cladoceran. *EcoScience* 12: 376–385.)

Figure 3.4 Histogram of length measurements from 1,651 mosquito fish. (Data from digital table 2)

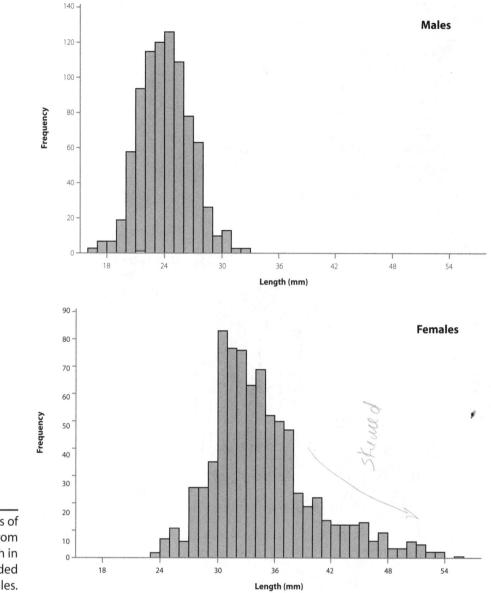

Figure 3.5 Histograms of length measurements from the mosquito fish in previous figure, divided into males and females.

gender has greater variation in body size? We will later calculate various descriptive statistics (chapter 4), and using graphics helps us to quickly spot if our answers make sense.

Most statistical programs for computers have graphing capabilities. After you have organized and plotted a couple of histograms by hand, you will appreciate this saving in labor! The graphs in figure 3.2 were generated by *Minitab*, using the raw data from table 3.3; similarly, the graphs in figures 3.4 and 3.5 used the raw data in digital appendix 2. Take time to copy these data into your statistical package and create a histogram. Explore the graphing capabilities and see if you can control the bin width to match these figures.

3.4 Cumulative Frequency Distributions

Occasionally, we would like to display data as cumulative frequencies, the number of observations in that class plus all lower classes. An example is shown for the mosquito fish data. Refer back to table 3.4 and confirm that the last column represents the running total of frequencies as you move down the table. A cumulative frequency distribution is graphed in two forms in figure 3.6 on the following page.

Such plots can be shown either as a histogram or as a smooth curve (sometimes called a "frequency polygon"). Notice that, as fish length increases, cumulative frequency continues to increase (as it must by definition). At some point the rate of increase slows and the curve starts to level out. is-Cumulative frequency graphs are also useful for revealing the diminishing returns from extensive sampling; e.g., "How many samples are enough to detect most of the species?" (fig. 3.7).

Cumulative frequencies can also be expressed as % and used to answer such questions as "What % of the population is smaller than or equal to size X?" This type of question is frequently raised with student scores from standardized exams (e.g., "What percentile was your score on the SAT [or ACT] exam?).

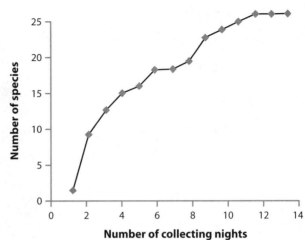

Figure 3.7 Example of cumulative frequency distributions: species accumulation curves for frogs from the Amazon. Source: Dahl, C., V. Novotny, J. Moravec, and S.J. Richards. 2009. *Journal of Biogeography* 36: 896–904.

Key Terms

bar graph	frequency distribution
bimodal	histogram
bins (classes)	skewed distribution
cumulative frequency distribution	frequency

Exercises

Answer the first three questions by hand. If you have a statistical program available, also construct the histograms using a computer. Do they look the same? Can you change the binning? Use a computer statistical package to solve the remaining questions.

3.1 Below are the heights (in cm) of a sample of 148 male humans. Construct a frequency distribution table and a histogram of these data. A class interval of 2 cm is suggested.

```
187  171  181  180  178  171  174  177  172  178  182
187  176  179  190  185  192  184  182  178  187  173
185  184  184  183  185  197  202  181  181  191  178
187  185  186  174  174  182  195  182  180  182  182
179  183  178  185  178  190  180  175  169  176  182
185  179  180  187  178  170  181  200  161  181  173
178  182  181  181  181  172  185  188  188  177  176
173  174  176  189  180  182  188  184  179  177  177
183  196  184  173  180  180  180  184  175  176  186
187  182  187  174  178  191  182  174  178  191  178
173  183  191  191  180  187  184  177  186  194  185
189  193  189  192  189  181  177  176  190  173  179
180  184  176  180  178  171  182  173  184  193  182
185  178  190  190  183
```

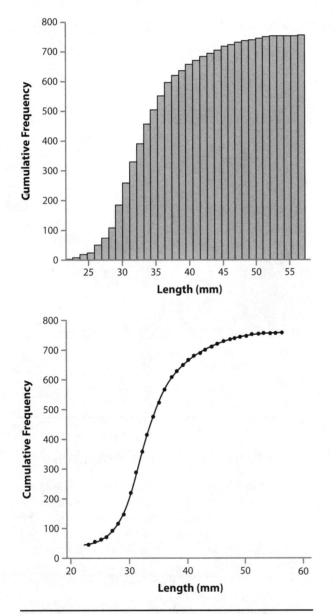

Figure 3.6 Cumulative frequency of 797 female mosquito fish lengths shown as a histogram (top) and as a frequency polygon (bottom).

3.2 Given below are the lengths (in millimeters) of a sample of 100 largemouth bass. Construct a frequency distribution table and histogram of these data.

210	325	285	402	350	240	409	330	295	325
241	383	361	355	200	432	130	114	170	135
371	307	207	175	177	261	166	376	216	152
347	322	387	233	284	394	297	321	281	66
90	115	250	201	175	320	370	312	370	320
175	201	250	115	95	70	289	312	322	258
188	192	350	200	199	180	190	180	200	200
349	192	189	260	320	432	456	331	418	357
304	316	336	368	415	370	336	315	305	420
310	397	193	394	199	338	296	312	269	203

3.3 Using the data on the lengths of bluegill sunfish in Digital Appendix 1 ("bluegill"), construct a histogram of this variable. A 10-mm class interval is recommended.

3.4 Using the data on the length of female mosquito fish in Digital Appendix 2 ("mosquito fish"), construct a histogram of this variable. A class interval of 4 mm is suggested.

3.5 Using the data on pulse rate in Digital Appendix 3 ("Student Data"), construct a separate histogram for the pulse rate of females and males and make sure you use the same scale for both. Are the distributions similar? If not, how do they differ?

3.6 Some species of chironomid larvae inhabit the leaves of pitcher plants. The number of larvae per leaf in a random sample of 197 leaves was as follows. Construct a bar graph of these data. What is the total number of chironomid larvae found in these 197 leaves?

Larvae/Leaf (x)	Number of Leaves Containing x Number of Larvae
0	10
1	15
2	27
3	18
4	38
5	57
6	22
7	5
8	2
9	3
10	0

3.7 Japanese eel (*Anguilla japonica*) are sometimes infested with the nematode parasite *Anguillicoloides crassus*. The number of adult worms found in 40 cultured eels is listed below. Construct a bar graph of these data. What is the total number of worms found in these 40 eels?

Worms/Eel (x)	Number of Eels Containing x Number of Worms
0	16
1	7
2	4
3	3
4	2
5	2
6	1
7	2
8	1
9	1
10	1

3.8 Construct a cumulative frequency distribution for the body length (in cm) data for 56 perch.

Measured Length (cm)	Number of Perch (f)
5–9	1
10–14	1
15–19	5
20–24	16
25–29	12
30–34	3
35–39	6
40–44	8
45–49	4

Data adapted from Heitlinger et al. 2009. Parasites and Vectors Vol. 48.

3.9 Construct a cumulative frequency distribution for the volume of water discharged (in cubic kilometers) by the Mississippi River over the 48 years from 1954 to 2001.

Measured Discharge (km^3)	Number of Years (f)
200–299	1
300–399	4
400–499	13
500–599	13
600–699	11
700–799	4
800–899	2

3.10 Locate an article in a field of your interest which contains numerical data. How are the data summarized for display in tables and/or figures? Can you find an example of a frequency distribution?

Descriptive Statistics
Measures of Central Tendency and Dispersion

If we measure some characteristic on a number of individuals in a population, we usually find that the measurement differs among individuals. Generally, there is a certain value around which our measurements tend to cluster. Statistics gives us ways to describe this central tendency and to describe the variation of individuals from this point. In this chapter we consider several ways to describe the central tendency and the variation around it.

4.1 Sample Statistics and Population Parameters

It would be quite unusual indeed to have all of a population available for measurement. Even if we did, we might not want to take all the trouble to measure so many things. Accordingly, we calculate such things as central tendencies and variation using a random sample from the population of interest and use them to estimate the population's values. These values in the population as a whole are called **parameters**,[1] and they are usually unknown. In a sample they are called **statistics** (plural; the singular is statistic). Later in this chapter, we calculate several statistics. Three of these are shown below with their accompanying population parameters. The true mean of a population is a parameter, and the mean of a sample taken from that population is a statistic. Parameters are conventionally symbolized by Greek letters and statistics by Roman letters, in this manner:

	parameter	statistic
mean	μ	$\bar{x}$
median	θ	M
variance	σ^2	s^2
standard deviation	σ	s

The Greek letters are: mu (μ), theta (θ), and sigma (σ).

4.2 Measures of Central Tendency

The **central tendency** of a sample (or a population) is the value around which measurements tend to cluster. Several statistics can be used to describe central tendency. Which is most appropriate depends on the type of variable we have and on the information we wish to convey.

4.2.1 The Mode

The **mode** is the most common value. For categorical variables, the only measure of central tendency is the mode. For example, in a group of 15 red marbles, 10 white marbles, and 5 blue marbles, the mode is red marbles. The mode can be applied to other variable types and is also used to describe the peak in a frequency distribution (fig. 4.1). As we saw in chapter 3, some continuous frequency distributions have more than one mode (fig. 3.4). Note that the peaks need not be the same height for the distribution to be multimodal.

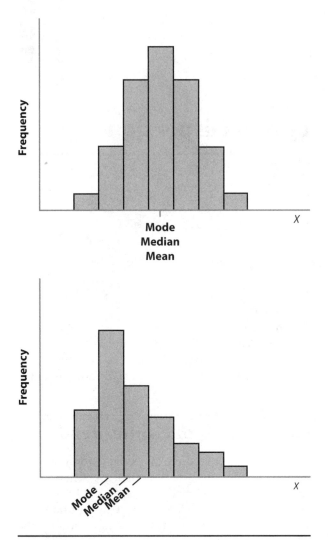

Figure 4.1 Symmetric and right-skewed frequency distributions. Adapted from Wonnacott and Wonnacott 1977.

4.2.2 The Median (θ, M)

The median is a useful measure of central tendency for ordinal (rank) data, as well as for measurements (below). Rank data indicate that the items can be arranged in order from smallest to largest. The **median** is the value that has an equal number of items above and below it. For example, suppose we have the following ordinal data on the conditions of 15 patients at Hopeless Hospital:

condition	near death	deteriorating	stable	improving
frequency	2	8	4	1

The median condition is "deteriorating." This result is made clearer if we collapse the frequency distribution into an ordered list of codes for the patient conditions, with the middle condition (median) underlined in bold:

N N D D D D D <u>D</u> D D S S S S I

For measurement variables, the median is the center of an ordered array of numbers. For example, in the measurements

2, 2, 2, 3, 3, 4, 4

The median is 3, since this value has an equal number of measurements above and below it. When there is an even number of observations, none can have an equal number above and below. In this case the median is the value halfway between the two central values. In the series

2, 2, 2, 3, 4, 4, 4, 4

the median is 3.5 (the average of 3 and 4). Note that in the series

1, 2, 2, 3, 4, 10, 100, 1000

the median is still 3.5, even though the latter three numbers have a great deal more spread and cover a much larger range than the first three numbers. The median is a very useful measure of central tendency for both rank and measurement data.

4.2.3 The Mean (μ, $\bar{x}$)

The **mean** is an arithmetic average and is the most common measurement of central tendency for measurement variables. If we measure all of the individuals in a population of interest and compute a mean based on these measurements, we have in fact obtained the true population mean (μ). On the other hand, if we measure only a randomly selected sample of the members of the population and use these measurements to compute a mean, we have obtained a sample mean, symbolized by $\bar{x}$. Ordinarily, we deal with samples, so we calculate sample means.

If x_1, x_2, x_3, ..., x_n are individual measurements from a sample of size n, their mean $\bar{x}$ is:

$$\bar{x} = \frac{\sum x}{n} \tag{4.1}$$

or, in words, the sum of all of the x values divided by the total number of x values. (By the way, the symbol Σ is the capitalized version of the Greek letter "sigma," and signifies summation.) The sample mean ($\bar{x}$) is an estimate of the true population mean (μ), which we ordinarily do not know. If we did measure all items in the population, μ would be calculated by the same formula, replacing $\bar{x}$ with μ.

4.2.4 Positions of Mean, Median, and Mode in Symmetric and Skewed Distributions

Suppose we collect a large number of observations and construct a frequency distribution. Some distributions are symmetric, with an equal spread of data on either side of the peak in frequency (mode). When this occurs, the mean, median, and mode all occur at the same value of x. Other distributions are skewed, with a peak on one side or the other of the distribution and a long "tail" to the other side. In a skewed distribution, the mode is again (by definition) the value of x where the peak in frequency occurs. The mean is located farther out toward the long tail and the median occurs between the mean and the mode. Examples of symmetric and skewed distributions are shown in figure 4.1.

Rare observations in the tail are sometimes called **outliers**. They heavily influence the mean. Let's imagine a common situation with a skewed distribution: incomes of workers. To simplify our problem, we will use a small sample.

Example 4.1

Annual incomes of the 9 employees of Redundancy Manufacturing (in $1000s):

15   17   21   21   24   27   27   31   225.

With so few measurements, the mode is not very meaningful. However, we can determine the mean and median. Because of the high salary of one person (presumably the boss), the mean is inflated to $45,333 (408/9), a value considerably larger than any employee other than the boss. Indeed, if we omitted this outlier, the mean salary of all the others is $22,875. The median salary for this company is $24,000. Notice that this value remains the same even if the boss took a large pay cut! This principle has political implications. Some economic indicators at the national level are inflated by using the mean, rather than the median, of income data from a population where most individuals are not millionaires but some earn far more.

4.2.5 Weighted Mean

The mean is typically calculated as the arithmetic average, as shown above. However, there are occasions when other types of means make more sense. Most common is the **weighted mean**, calculated from data arranged in a frequency distribution as follows:

$$\bar{x} = \frac{\sum fx}{\sum f} \qquad (4.2)$$

where Σfx is the sum of products between values of x and their frequencies (weights), and Σf is the sum of the frequencies, also equal to n, the total sample size.

Example 4.2

Most undergraduates are concerned with their grade point averages, which are simply weighted means! For example, let's consider Sue Sigma, who has taken 50 credits so far and passed 46 of these. When all her grades are compiled into a frequency distribution (look familiar?), we get:

grade	grade value x	number of credits attempted f	grade points fx
A	4	12	48
B	3	18	54
C	2	12	24
D	1	4	4
F	0	4	0
		$\Sigma f = 50$	$\Sigma fx = 130$

$$\bar{x} = \frac{\sum fx}{\sum f} = \frac{130}{50} = 2.60$$

Weighted means are calculated when different measures have different levels of importance (weights). In the GPA example, the number of credits represents the weight. As another example, consider a problem from environmental toxicology. Suppose we had PCB measurements (x) of tissues taken from different individual lake trout, which we had previously weighed. In order to estimate the overall PCB levels in lake trout, we would determine a weighted mean, using the fish weight as the "weight" (f above).

Weighted means are also useful to determine an average from data that already are arranged in a frequency distribution. For instance, consider the data on mosquito fish lengths in the last chapter (table 3.4). The data are tabulated for us. The weighted mean would then be $\Sigma fx \div \Sigma f = 3744 \div 172 = 21.8$ mm. Notice that this value is near the center of the frequency distribution (fig. 3.2). Calculating a mean from such a frequency distribution is often useful to get additional information from summary data in a publication. Of course, if we had the original data at our disposal, we could simply calculate the sample mean and have no need to calculate the mean from the frequency distribution.

4.2.6 Other Measures of Location

Recall that the median is a measure of central location. In an ordered array, the median divides the top half of observations from the bottom half. There are other locations we could mark. **Percentiles** divide up an ordered array into 100 equal slices. Having taken standardized tests, most of us are familiar with percentiles. If you scored in the 90th percentile in the verbal test, this means that 90% of the people taking the test scored below you.

As another example, consider the cumulative frequency of 797 female mosquito fish (fig. 3.6). Figure 4.2 takes that same distribution, converting the

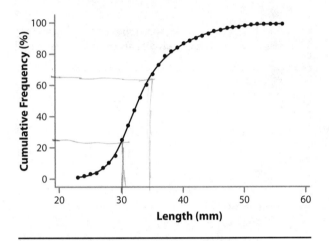

Figure 4.2 Cumulative frequency (as %) for 797 female mosquito fish, shown against length.

Example 4.3

Consider an imaginary example from Wisconsin. A farmer has 14 cows, from which he determined milk yields. Milk yield is measured in pounds per day and the cows' measurements are sorted in ascending order:

19 23 26 30 32 34 37 37 39 41 44 44 46 55

---

frequencies to %. To interpret the graph, we can say that 24.6% of the population are less than or equal to 30 mm and 87.2% are less than or equal to 87.2 mm. Looked at another way, we can see that the median (50th percentile) is a value close to 33 mm and the 25th and 75th percentiles are at approximately 30 and 36.5 mm respectively. The 25th, 50th, and 75th percentiles are also called **quartiles**, since they divide the distribution into quarters. The **first quartile** (Q1) divides the bottom quarter from the top three quarters and is the same as the 25th percentile. The median is also called the **second quartile** (and 50th percentile), as it divides the bottom two quarters from the top two quarters. The **third quartile** (Q3) divides the top quarter from the bottom three quarters.

Using equation 4.1, we calculate the mean milk yield to be 36.2 pounds per day. Since the number of measurements is even, the median is the average between the 7th and 8th positions (both 37). A "quick and dirty" way to find the first quartile (Q1) is to find the number that divides the lower half of observations in half. In this example, that number occurs at about the 4th position, where the value is 30. By a similar method, the third quartile (Q3) has a value of 44. This should satisfy most of us for visualizing the quartiles. In everyday work, we typically obtain exact results with computer software, as shown for *Minitab* below (table 4.1).

In the table, recognize the values for the sample size (14 cows), mean, minimum, median, and maximum, and confirm that they are the same as we determined by hand or by inspecting the ordered list (example 4.3). The quartiles are also shown. Notice that our quick-and-dirty estimate of Q1 (30) does not exactly match the *Minitab* result (29). This is because we used an approximation and *Minitab* determines the exact value. The exact approach requires some interpolation:

Table 4.1   Descriptive Statistics from *Minitab*

A) Milk yields from 14 Wisconsin cows (example 4.3)

Descriptive Statistics: MilkYield

Variable	N	N*	Mean	SE Mean	StDev	Minimum	Q1	Median	Q3
MilkYield	14	0	36.21	2.61	9.76	19.00	29.00	37.00	44.00

Variable	Maximum
MilkYield	55.00

B) Lengths of 172 male mosquito fish (example 3.2)

Descriptive Statistics: MosquitoFishL

Variable	N	N*	Mean	SE Mean	StDev	Minimum	Q1	Median	Q3
Mosquito FishL	172	0	21.831	0.155	2.026	17.000	20.000	22.000	23.000

Variable	Maximum
Mosquito FishL	30.000

1. Calculate $(n + 1)/4$.
2. If $(n + 1)/4$ is an integer (whole number), then Q1 is the value of this integer.
3. If $(n + 1)/4$ is not an integer, then Q1 is a value between two observations and its location between the numbers is proportional to the value of the decimal answer. For the cows data, $(14 + 1)/4 = 3.75$. Q1 is thus ¾ of the distance between the 3rd and 4th numbers, a value of 29. (This was figured as: $26 + 0.75(30 - 26) = 26 + 3 = 29$.) To find the third quartile (Q3), first calculate $3 × (n + 1)/4$ and use the same approach as above. For the cows, $3 × (14 + 1)/4 = 11.25$. Q3 is thus ¼ of the way between the 11th and 12th numbers, in this case both 44. So $Q3 = 44 + 0.25 (44 - 44) = 44$.

We have determined a variety of statistics for central tendency and other locations in a sample distribution. We are now ready to consider variation, the stuff that makes statistics necessary and interesting!

4.3 Measures of Dispersion

It is unlikely that individuals of a biological population will all be exactly alike in any variable that one cares to measure. We usually find that individuals differ from one another with respect to such things as length, weight, number of warts, and so on. As variation is the usual state of the natural world, it is quite important to measure it. Controlling blood pressure with medication allows not only a reduction in average diastolic blood pressure, but also its swings from high to low. A drunken driver on the highway is dangerous to others not so much because of their mean position, but because of their variability in position—weaving down the highway! For all these things, there is some value about which individual measurements tend to cluster. For discrete and continuous measurements, this value is usually the arithmetic mean, and when measuring with an ordinal scale this value is the median.

Consider the data on length of 172 male mosquito fish. Refer back to chapter 3 to see the raw data (table 3.3) and histogram from these data (fig. 3.2). The mean for this sample is 21.83 mm. Note that the most frequently occurring size class is made up of individuals whose size clusters about the mean value. Note also that there are progressively fewer individuals in each size class as we move from the mean toward either smaller or larger individuals. Frequently, we need to measure this spread of the individual observations around the mean. The terms spread and **dispersion** both refer to the same thing, the variation in observations around the mean (or median).

4.3.1 The Range

The **range** is the difference between the largest and smallest items in the sample. The range is expressed in the same units as the original measurement. For example, the smallest measurement in the data on male mosquito fish length (table 3.3) is 17 mm, and the largest is 30 mm. The range is therefore 30 mm – 17 mm, or 13 mm. Similarly, for the Wisconsin cow data above, the range is 55 – 19, or 36 pounds of milk per day.

The range is strongly influenced by even a single extreme value. Suppose, for instance, that a single cow produced 162 pounds per day. The range would now be 143 pounds per day, even though the spread among the other cows was the same. The range is thus limited as a description of the amount by which any individual measurement is likely to vary from the mean. Nevertheless, reporting the maximum and minimum values is a common description of environmental and physiological data, because of the tolerance limits of organisms. For instance, your body temperature may usually vary little from 98.6°F, but one instance of extremely high temperature, say over 107°F, can be fatal. When reporting the range, the best approach is simply to report both the minimum and maximum and the reader can do what they want with these numbers.

4.3.2 Interquartile Range

Because of the extreme effect of outliers on the range, another measure of dispersion is often presented. The **interquartile range** is the difference between the third and first quartiles and represents the spread of the middle 50% of the (ordered) values. For the cow data, the interquartile range is about 44 – 29, or 15 pounds per day. In other words, the middle half of cows showed a range of 15 pounds per day. The interquartile range often appears on a graph. The **Box-Whisker Plot** (usually just called the Box plot) illustrates the locations of the median, interquartile range, and range, and is often used to compare these characteristics in different groups (fig. 4.3). For this example, we have added another hypothetical sample, one from Missouri, which has some very productive cows!

The median is shown as the horizontal line inside the box and the first and third quartiles are at the boundaries of the box. The vertical lines (called "whiskers") represent the range of most values.

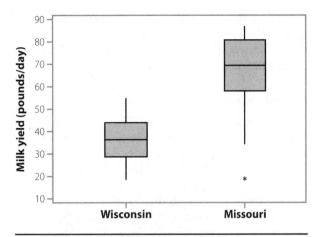

Figure 4.3 Box plots of milk yields data from two locations, as generated by *Minitab*. [see text for explanation]

Specifically, the top whisker extends up from Q3 to the maximum value that is still within the limit of Q3 + 1.5 × (Q3 − Q1) and the bottom whisker extends down to the minimum value which is still within the limit Q1 − 1.5 × (Q3 − Q1). This corresponds approximately to the 10th and 90th percentiles. Outliers show up as an *. (One outlier is shown for the Missouri data.) Since computer software can readily generate such plots, most users don't need to do these calculations.

Box plots are becoming more common in the scientific literature, particularly for measurement variables that are highly skewed in their distributions. For instance, densities of bacteria in the environment can vary over a large range, with low values on most days and extremely high values on others.

4.3.3 Standard Deviation (σ, s)

Next, let's consider a widely used measure of variation that takes into account all of the observations in a distribution. The **standard deviation** is, in effect, a measure of the average amount by which each observation in a series of observations differs from the mean. We will use an imaginary set of data to illustrate the calculation and meaning of the standard deviation. This calculation is only possible with discrete and continuous measurement variables. The true population standard deviation, denoted by the Greek symbol sigma (σ), is usually unknown and must be estimated from our sample, calculating the sample standard deviation (s).

Suppose we collect a sample of 8 measurements: 1, 2, 2, 3, 3, 4, 4, 5. The sample mean for these 8 measurements is:

$$\bar{x} = \frac{1+2+2+3+3+4+4+5}{8} = \frac{24}{8} = 3.0$$

We could compute an average deviation by first subtracting each measurement (variant) from the mean, then summing these values and dividing by the number of observations. However, we would find that the sum of these **deviations**, symbolized by d, would equal zero (see table 4.2 below), and the average deviation of the observations from the mean would therefore be zero. In fact, we would find that the sum of the deviates around any mean is always zero! Squaring each deviate eliminates the negative signs. When the squared deviates are summed, the resulting value is called the **sum of squares**, which we will abbreviate as SS.

Caution
The term "sum of squares" (abbreviated SS) is commonly used in this book and refers to the sum of squared deviations from a mean. The sum of squares does not mean simply the sum of the squared observations (Σx^2 below). Confusing these two quantities is a very common mistake by students.

The sum of squares (SS) is obtained by squaring the difference (deviation) between each observation and the mean of all of the observations, and then summing the squared differences, or

$$SS = \Sigma(x - \bar{x})^2 \qquad (4.3)$$

As we shall see later, the SS is a very important quantity in many statistical procedures. In our example, the sum of squares is 12 (table 4.2).

The procedure of subtracting each observation from the mean, squaring the difference, and summing these squared deviates is computationally tedious if more than a few observations are involved. A much more efficient method of computing the sum of squares is by formula 4.4:

$$SS = \Sigma x^2 - \frac{(\Sigma x)^2}{n} \qquad (4.4)$$

The first term in the equation, x^2, instructs one to square each observation and then sum the results. The second term, $(\Sigma x)^2/n$, instructs one to sum the observations, square the sum, and then divide that result by the number of observations.

If the sum of squares is divided by the number of observations minus one, another important value, called the **variance**, is obtained:

$$s^2 = \text{variance} = \frac{SS}{n-1} \qquad (4.5)$$

The variance represents the average of the squared deviations in the sample. If the original measurements were in mm, the units of the variance would be mm². Ordinarily, we desire a statistic with the same scale and units as our original measurements. To rectify this condition, we need only take the square root of the variance. This value is called the **standard deviation**. The standard deviation has the same units as the original measurements.

$$s = \text{standard deviation} = \sqrt{\frac{SS}{n-1}} = \sqrt{\text{variance}} \quad \textbf{(4.6)}$$

In our example above, the standard deviation is 1.309. Work through this example, summarized in table 4.2, to make certain that you thoroughly comprehend these concepts.

Table 4.2 Calculation of Mean, Variance, and Standard Deviation from a Sample

The sample mean equals 3.0 and sample size is 8.

x	$d = x - \bar{x}$	d^2
1	−2	4
2	−1	1
2	−1	1
3	0	0
3	0	0
4	1	1
4	1	1
5	2	4
	$\Sigma = 0$	$SS = 12$

$$s^2 = \text{variance} = 12/7 = 1.7143$$

$$s = \text{standard deviation} = \sqrt{1.7143} = 1.3093 \approx 1.3$$

Notice that we have carried all figures in this example and rounded only at the end. We round these sample statistics to one more significant figure than the original measurements. So, after rounding, the standard deviation should be reported as 1.3 and the mean as 3.0.

Caution

Beware of rounding errors. When doing calculations, always carry all the figures in your calculator memories. Round only at the final step. Rounding errors can sometimes be quite significant! Clever embezzlers have occasionally taken advantage of rounding errors, by skimming money from multiple transactions in banks.

Calculate the standard deviation for the cow milk yield example with your hand calculator (or spreadsheet) and confirm the standard deviation reported in table 4.1. Review the other statistics

shown in the table. Most of the statistics should now look familiar, with a few exceptions. "N*" is the number of missing observations. "SE Mean" is the "standard error of the mean," a measure of variation which we consider in a later chapter.

4.3.4 Coefficient of Variation

There are times when we wish to re-express the standard deviation in such a way as to allow a comparison between groups that are very different in some way. Suppose, for instance, that we wanted to compare the variability in body weights between mice and elephants. Simply comparing their standard deviations just doesn't make sense; after all, these animals would have been measured on different scales. Instead, we'll standardize their standard deviation relative to their mean body weight. To do so we compute the **coefficient of variation** as:

$$CV = \frac{s}{\bar{x}} * 100\% \quad \textbf{(4.7)}$$

The following hypothetical example will illustrate:

Example 4.4

	Body weight (kg)	
	mice	elephants
$\bar{x}$	0.0125	1,240
s	0.0072	625
CV (%)	57.6	50.4

The example shows that, after standardizing to their mean body weights, variability in body weights is quite similar for mice and elephants.

Key Terms

Box-Whisker Plot
central tendency
coefficient of variation
dispersion
interquartile range
mean
median
mode
outlier
parameter
range
standard deviation
sum of squares (SS)
variance
weighted mean

Exercises

4.1 For the following sample, compute the following statistics by hand using a calculator: mean, median, range, interquartile range, variance, standard deviation, coefficient of variation. What is the sample size? Show all steps in your solutions. Note that you will need to rewrite the numbers in order from smallest to largest. Finally, use a statistical package to generate the same descriptive statistics you calculated by hand. Do your answers agree?

2 5 3 7 8 3 9 3 10 4 7 4 6 11 9
9 11 5 7 3 8 9 2 1 3 8 3 8 9 3

4.2 For the following sample, compute the same statistics as in exercise 4.1 and use a statistical package to check your answers.

23 43 12 56 43 23 56 43 23 32 12 14 15

4.3 A random sample of 42 belted kingfishers was collected from various locations in North America and their culmen (bill) lengths were measured in mm:

48.1 50.8 48.8 56.8 57.7 47.0
56.8 60.2 55.8 59.2 52.5 50.4
48.0 57.1 51.8 52.3 47.8 58.0
53.4 55.2 51.0 59.3 61.5 61.2
57.8 50.1 56.0 56.5 55.8 56.5
56.3 59.8 61.8 56.2 57.5 59.3
62.4 61.1 59.9 55.6 56.8 59.2

Compute the same statistics as in exercise 4.1 and use a statistical package to check your answers.

4.4 Using the frequency distribution table below for the carapace length of 47 painted turtles, compute a weighted mean (eq. 4.2). Compare to the actual sample mean of 123.6 mm.

Class Mark (x, mm)	f
95	4
105	10
115	9
125	6
135	9
145	3
155	5
165	1

[handwritten annotations: 95 × 4 = 380; 105 × 10 = 1050; 115 × 9 = 1035; 125 × 6 = 750; 135 × 9 = 1215; 145 × 3 = 435; 155 × 5 = 775; 165 × 1 = 165; 5805/47 = 123.]

Data adapted from Jolicoeur and Mosimann 1960. *Growth* 24: 339–354.

4.5 Data on the brood-size of great tits was collected from 1958–1963. Using the frequency data below for the brood-size, compute the weighted mean for the brood size over the 6 years (eq. 4.2).

Brood size (x)	f
1	5
2	10
3	32
4	61
5	81
6	95
7	80
8	88
9	72
10	67
11	61
12	41
13	24
14	1
15	1
16	2

Data adapted from Perrins 1965. *Journal of Animal Ecology* 34: 601–647.

4.6 Using the frequency data from exercise 2.3, compute the mean number of ant lion pits/quadrat for this sample. Save your calculated value for later use.

4.7 Using the ordinal data from exercise 2.4, select and compute an appropriate measure of central tendency.

4.8 The antipredator behavior by Willow Ptarmigan, a bird common in the arctic tundra, was analyzed during the incubation period. Antipredator behavior of mothers was ranked according to their proximity and conspicuousness to predators, with 1 being the least risky and 5 being the most risky. For the ordinal data below, select and compute an appropriate measure of central tendency.

Defense Rank	f
1	4
2	9
3	4
4	6
5	1

Data adapted from Martin and Horn 1993. *OrnisScandinavica* 24: 261–266.

4.9 Using samples of 10, 20, and 30 bluegill-sunfish standard lengths from exercise 1.1, compute the mean and standard deviation of each sample using a computer. Consider the measurements in digital appendix 1 to be the entire population of interest, with a population mean of 120.03 mm and a population standard deviation of 41.95 mm. Are the sample means and standard deviations of your samples the same as the population mean and standard deviation? Why do you suppose that there might be a difference in the sample values and the population values? Save your calculated values for later use.

4.10 Using the samples of 10, 20, and 30 female mosquito fish lengths from exercise 3.5, compute the mean and standard deviation for these samples using a computer. Consider the measurements in digital appendix 1 to be the entire population of interest, with a population mean of 34.29 mm and a population standard deviation of 5.49 mm. Are the means and standard deviations of your samples the same as the population mean and standard deviation? How do you account for the difference? Save your calculated values for later use.

Note

[1] Sometimes the term *parameter* is used in another way, which we view as incorrect. Limnologists (and perhaps other scientists) sometimes refer to measuring "parameters" such as pH, phosphorus, chlorophyll, etc. It is better to use the term "variable" for these properties and restrict "parameter" to its exact statistical meaning. Of course the term "sample" can also be ambiguous; a sample may represent a single collection of something (water, blood, etc.) or, in the statistical sense, a collection of measurements for a variable of interest.

chapter five

Probability and Discrete Probability Distributions

Most events in life are uncertain. Will it rain today? Will the batter strike out? Will the phone ring tonight? **Probability**[1] is the chance that a particular event will occur in an uncertain world. The higher the probability, the more certain we can be that the event will occur. When the probability is 0, the event is impossible; and when the probability is 1, the event is certain. Probability generates expectations that support informal decisions in such disparate fields as business, sports, and gambling. (Indeed, the formal rules of probability were developed in the 1600s by mathematicians employed by gamblers; now they are employed by insurance companies!) Probability rules are used in biology to predict the

risk of genetic disease if we know a pattern of inheritance, and the risk of other diseases if we know some test outcome (using Bayesian statistics). Probability is a foundation of statistical inference (hypothesis testing), which we explore extensively later in this book.

In the current chapter we consider some of the basic rules of probability and some important discrete probability distributions. In chapter 8, we will examine another important probability distribution called the normal distribution.

5.1 Probability

We can find the probability of an event in a couple of different ways. One is based on a theoretical consideration (also called "**classical probability**"). For example, dice have 6 faces, each bearing a number of 1 through 6. Given that a die[2] is "fair" (that is,

Figure 5.1 Each die has 6 faces and thus 6 possible outcomes, and a coin has 2 faces and thus 2 possible outcomes.

equally as likely to land with one face up as any other), then the probability of obtaining any particular number on the toss of a single die is 1/6 or 0.1667. Flipping coins, drawing cards from a deck, and rolling dice are all governed by the rules of probability, based on the structure of the game. Similarly, for genes following simple rules of Mendelian inheritance, we can predict the proportion of individuals expressing a recessive phenotype following a monohybrid cross. In classical probability we know the "rules of the game" and thus can make predictions using some basic tricks (rules), which we'll discover in this chapter.

A second important way of determining probability is based on some prior knowledge of the relative frequency of the event in the population of interest. This relative frequency is called **empirical probability**. For example, the occurrence of children born with Down's syndrome (a form of mental retardation) is well known from birth records, so its incidence (relative frequency) can be calculated for different groups. From this information, we know that the incidence rises sharply in mothers after age 35. Doctors can thus give their patients an assessment of risk. Note that, in this case, we can arrive at this probability even though we don't know the precise mechanism. For both classical and empirical probability, probability was calculated from a ratio. This is the first of our probability rules, called the division rule.

The Division Rule

The probability of an event is the number of ways that an event can occur divided by the total number of events that may occur.[3] In other words:

$$p = s/n \qquad (5.1)$$

If we toss a coin, there are 2 possible outcomes—a head or a tail. Since the coin has an equal number of heads and tails (one of each), on any given toss there is an equal chance of obtaining a head or a tail. Furthermore, the 2 possible outcomes are **mutually exclusive**, which means that both events cannot occur simultaneously. Thus, using the division rule, the probability that any toss of the coin will produce a head is 1 (the outcome of interest) divided by 2 (the number of possible outcomes), or 1/2 (0.5). The probability that any toss will result in a tail is, of course, also 1/2 (0.5) and is deduced in the same manner as above.

For the coin tossing example, we may compute the probability of a tail in this manner: let p = the probability of a head and q = the probability of a tail.

The Subtraction Rule

The probability of an event *not* occurring is equal to one minus the probability of that event occurring. In other words:

$$q = 1 - p \qquad (5.2)$$

tail. Since the probability that the outcome will be one or the other (a head or a tail) is 2/2 = 1.00, then

$p + q = 1$, and
$q = 1 - p$, or
$q = 1 - 0.5 = 0.5$

In the case of the dice, there was one way that a particular number could occur out of 6 possible occurrences. Let's say we let p = the probability of rolling a "six." Then $p = 1/6$. What is the probability of rolling anything else (q)? We could count all the possibilities (five of them) and then compute q as 5/6. Or we could do as above and subtract: $q = 1 - 1/6 = 5/6$. In other words, we have computed the probability of *not* rolling a "six."

A **probability distribution** is a listing, in the form of a table or graph, of the probabilities associated with each possible outcome. A probability distribution that describes a single toss of the coin is:

Outcome	Probability
head	0.5
tail	0.5

Note that we need not toss any coins to arrive at this probability distribution. It is a theoretical distribution that we derive from our knowledge of how flipped coins behave and from the laws of probability. Probability distributions are useful in allowing us to predict what should tend to happen in the real world. If we tossed a coin 100 times, we would expect to obtain approximately 50 heads and 50 tails if our assumptions about coin behavior under these conditions are correct.

If we toss a die, the probability of obtaining a "one" is 1/6 (0.1667), the probability of obtaining a "two" is also 1/6 (0.1667), and so on, for each of the 6 possible outcomes. The probability distribution for the outcomes of a single toss of a die is shown below in a table.

Outcome [x]	1	2	3	4	5	6
Probability [$p(x)$]	1/6	1/6	1/6	1/6	1/6	1/6

Below we'll consider repeating such experiments over more than one trial. We are interested in the probabilities associated with all the possible occurrences of the event in question. Several probability distributions describe the behavior of discrete variables. Two of these, the Binomial distribution and the Poisson distribution, are considered in this chapter. Before describing these distributions, we should consider how to count all possible outcomes from a trial. We will also examine two more fundamental rules of probability.

5.2 Counting Possibilities

We can count possibilities in two ways. One way is to see how many different sequences are possible. For example, we might consider four letters (let's say *a, p, s, t*) and ask how many different ways we could arrange them in a line to form hypothetical words. You could list all the possibilities (there are 24) or recognize that this is asking for **permutations**. The number of permutations is simply calculated as the factorial of the number of interest (*n!*). For the words example, 4! = 4*3*2*1 = 24. (Recall the rules for factorials—see footnote.)[4]

The other way to count possibilities is to count the number of groups, ignoring sequence. The number of possible groups is called the number of **combinations**. The formula for the number of combinations of *k* objects taken *x* at a time ("*k* choose *x*") is given by equation 5.3.

$$_kC_x = \binom{k}{x} = \frac{k!}{x!(k-x)!} \qquad \text{(5.3)}$$

Suppose I have a cage of 5 hamsters and wish to select 3 of them. How many different groups are possible? Using the equation above, "5 choose 3" would be equal to 5! / 3! (5 – 3)! = 5*4*3! / (3!*2*1) = 10. We could use a similar approach to determine the number of combinations of cards ("card hands") that are possible when drawing from a deck of cards. Can you figure the number of 5-card poker hands that are possible from a deck of 52 cards? This is quite a large number.

More relevant to our purposes here is the fact that the Binomial distribution (section 5.4 below) uses the number of combinations as part of the formula for computing each of the Binomial probabilities. Before introducing the Binomial distribution, we need to look at two more probability rules.

5.3 More Probability Rules

Consider the possible events that can take place when you toss 2 coins simultaneously. The result can be 2 heads, 1 head and 1 tail, or 2 tails. What would the probability distribution for this situation look like? In other words, what is the probability of each of these 3 possible outcomes? The probability

that one coin will be a head is 0.5 (1/2), and the probability that the other coin will be a head is also 0.5. Note that these 2 outcomes are independent of each other—the way that one coin lands has no effect on the way that the other coin lands. The probability that both will be heads is the product of their individual probabilities, which is 0.5 × 0.5, or 0.25.

The Multiplication Rule (the "And" Rule)

The probability that independent events will occur simultaneously (that event A and event B will both occur) is the product of the probabilities that the events will occur individually. Stated another way: if A and B are independent events with probabilities *p*(A) and *p*(B), the probability of the joint occurrence of both A and B is

$$p(A \text{ and } B) = p(A) \times p(B) \qquad \text{(5.4)}$$

Two events are **independent** when the occurrence of the first event has no effect on the probability of the second event occurring. If events are not independent, then we have **conditional probability**, where the probability of a second event occurring depends on whether or not the first event occurred. Here, we must use a modification of the multiplication rule:

$$p(A \text{ and } B) = p(A) \times p(B|A) \qquad \text{(5.5)}$$

The term **B|A** is read "B given that A occurred" or just "B given A." An example will help. Imagine you are dealing from a deck of cards. There are 4 kings in the deck of 52 cards total. If you were to deal 2 cards *without replacement*, what is the chance of getting 2 kings? The probability would be:

$$p(2 \text{ kings}) = p(\text{king on 1st}) \times$$
$$p(\text{king on 2nd} | \text{king on 1st})$$

Using the division rule, the *p*(king on 1st) = 4/52 = 1/13. Now the deck has changed. If you already drew a king on the 1st draw (a success), then there remain 3 kings in a deck of 51 cards. So, the *p*(king on 2nd | king on 1st) = 3/51 = 1/17. Finally, the joint probability would be: (1/13)*(1/17) = 1/221.

Try one more example from the same deck of cards. There are 4 kings and 4 queens in the deck of 52 cards. If you were to deal 2 cards *without replacement*, what is the chance of getting a king followed by a queen? As before, *p*(king on 1st) = 4/52 = 1/13. Now that you already drew a king, there are still 4 queens among 51 total cards remaining. So the joint probability: *p*(king on 1st and queen on 2nd) is 1/13 × 4/51 = 4/663. (Note that we could use the same approach to obtain the chance of getting 1 queen followed by a king.)

$$\frac{4}{52} \times \frac{4}{51} =$$

In the biological world, whether sampling is effectively with replacement or without replacement depends on the situation (sometimes called the sampling universe). If a pond has a very large population of frogs and you are sampling for mutants, multiple individuals can be removed and have very little impact on the remaining population. So, in this situation, you can treat the sampling as independent. If, on the other hand, the population is more restricted—say 50 individuals—then, the chance of getting mutant individuals will be effected by the results of previous draws from the population, analogous to our deck of cards.

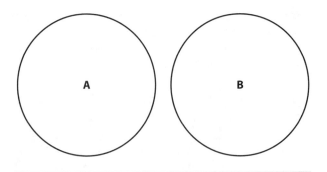

Figure 5.2   A Venn diagram, illustrating two mutually-exclusive events A and B.

The Addition Rule (the "Or" Rule)

The probability that at least 1 of 2 or more alternative outcomes will occur is the sum of the individual probabilities that the events will occur. To be more precise, if A and B are mutually exclusive events with probabilities $p(A)$ and $p(B)$, the probability that either event A or event B will occur is:

$$p(A \text{ or } B) = p(A) + p(B) \qquad (5.6)$$

For example, by the multiplication rule we determine that the probability of 2 tails is 0.25 (0.5 × 0.5). Now, what is the probability that we get one head and one tail? There are 2 ways that the occurrence of one head and one tail can result. The first coin can be a tail ($p = 0.5$) and the second coin can be a head ($p = 0.5$), and the probability of this result is 0.5 × 0.5, or 0.25. On the other hand, the first coin can be a head ($p = 0.5$) and the second coin can be a tail ($p = 0.5$). The probability of this result is, as before, 0.25. Since the probability of a head and a tail is 0.25 and the probability of a tail and a head is 0.25, the probability of this event happening in either of these 2 ways is 0.25 + 0.25 = 0.5.

Notice the condition in our definition. We assumed that events A and B are mutually exclusive events. In our coin flip example above, we cannot simultaneously obtain HT and TH; each coin will land either one way or the other. In contrast, when two events are not mutually exclusive, one event can occur at the same time as the other event. For example, with a deck of cards it would be possible to draw both a king and a heart simultaneously. Similarly, a heart patient could have both high blood pressure and high cholesterol simultaneously.

A visual aid for thinking about such events is the Venn diagram (figs. 5.2 and 5.3). The first diagram depicts mutually exclusive events: the circles indicating the occurrences of events A and B occurring do not overlap. In contrast, the second diagram

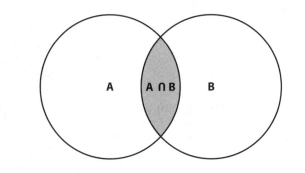

Figure 5.3   A Venn diagram, illustrating the intersection of two events A and B that are not mutually exclusive.

illustrates overlap between the two circles (fig. 5.3) indicating that they are not mutually exclusive. The expression A∩B represents the intersection between events A and B.

When events are not mutually exclusive we must use a general version of the addition rule. The probability that at least 1 of 2 alternative outcomes will occur is the sum of the probability of each event minus the probability of their joint occurrence:

$$p(A \text{ or } B) = p(A) + p(B) - p(A \text{ and } B) \qquad (5.7)$$

Using the Venn diagram (fig. 5.3) to illustrate probabilities, the intersection represents the joint occurrence; if we weren't subtracting this area we would be counting it twice. As an example of this general addition rule, consider again the case of the cards. If the probability of a king is 4/52, of a heart is 13/52, and of a king of hearts is 1/52, then the probability of being a king *or* a heart is 4/52 + 13/52 − 1/52 = 16/52 or 4/13.

5.4 The Binomial Distribution

Many objects or events belong to one of two mutually exclusive categories. Heads or tails, male or female, present or absent, and alive or dead are obvious examples. Even when there are more than two mutually exclusive categories, it is sometimes possible and useful to "lump" various categories into two by considering the object or event of interest as one category and everything else as the other category. Thus, "red" and "not red" are two mutually exclusive categories, even though "not red" can consist of blue, green, yellow, or any of a large number of colors that are not red. In the toss of a die, we might designate "1" as one possible outcome and "not 1" (any result except 1) as the other alternative.

The **binomial probability distribution** describes the probabilities associated with all possible outcomes of events when the events have the following properties:

1. The event (the toss of a coin, for example), sometimes called a trial, occurs a specified number of times, designated as k.

2. Each time the event occurs, there are 2 mutually exclusive outcomes (a head or a tail). One of these outcomes is specified as the outcome of interest (sometimes arbitrarily called a "success"). The probability of this outcome is designated as p. The probability of the other outcome (a "failure" in statistical jargon) is designated as q. Since there are only 2 possible outcomes, $p + q = 1$ and $q = 1 - p$. $p = \text{succes}$

3. The events are independent (see 5.3 above).

4. The number of times that the outcome of interest occurs in k events is designated as x. The binomial probability distribution gives the probabilities of x, symbolized by $p(x)$, for values of x from zero to k.

Recall the coin tossing example from above. It might help to visualize this situation in a table, as shown below. The rows represent the possible outcomes of one coin, and the columns represent the possible outcomes of the other coin. We compute the probabilities of the possible joint occurrences of the 2 coins simply by multiplying the columns by the rows. (This approach may remind you of the Punnett square treatment in genetics. That is because it is the same concept!)

The Other Coin

One Coin	H ($p = 0.5$)	T ($p = 0.5$)
H ($p = 0.5$)	HH ($p = 0.5 \times 0.5$)	HT ($p = 0.5 \times 0.5$)
T ($p = 0.5$)	TH ($p = 0.5 \times 0.5$)	TT ($p = 0.5 \times 0.5$)

We may deduce these probabilities in an intuitive manner, as we did above, or we may simply expand the binomial expression:

$$(p + q)^k = 1 \qquad (5.8)$$

in which p is the probability of one outcome of interest (heads), q is the probability of the other outcome (tails), and k is the number of events. In our example, p is 0.5 (the probability that any coin toss will result in a head), q is $1 - p$ ($1 - 0.5 = 0.5$), and k is 2 (the number of coins tossed). Expanding the binomial expression above gives

$$p^2 + 2pq + q^2 = 1, \text{ or}$$
$$0.25 + 0.50 + 0.25 = 1$$

Each term in the expanded binomial expression gives the probability of one of the possible outcomes. Notice that these probabilities are the same as those we deduced earlier with a more intuitive approach. Notice also that the sum of all these probabilities is 1.

A bar graph of this probability distribution is shown in figure 5.4.

What are the probabilities of the various outcomes when 3 coins are tossed ($k = 3$)? The possible outcomes are: 3 heads, 2 heads and 1 tail, 1 head and 2 tails, and 3 tails. As before, p (the probability that any coin will be a head) is 0.5, and q is $1 - p$, or 0.5. Using the same logic as above, the probability of 3 heads would be $0.5 \times 0.5 \times 0.5$ (= 0.5^3). Similarly, the outcome of 2 heads (and a tail) can occur three different ways (HHT, HTH, and THH); thus the probability of 2 heads is $3 \times 0.5 \times 0.5 \times 0.5$ (= $3 \times 0.5^3 = 0.375$). Expanding the binomial expression and substituting the probabilities for p and q:

$$(p + q)^3 = 1$$
gives
$$p^3 + 3p^2q + 3pq^2 + q^3 = 1, \text{ or}$$
$$0.125 + 0.375 + 0.375 + 0.125 = 1$$

These then are the probabilities associated with each of the 4 possible outcomes (0.125 for 3

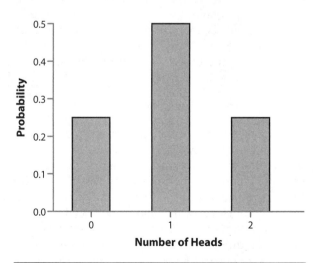

Figure 5.4   Binomial probability distribution for $k = 2$ and $p = 0.5$

heads, 0.375 for 2 heads and 1 tail, and so on). This probability distribution is shown as a bar graph in figure 5.5.

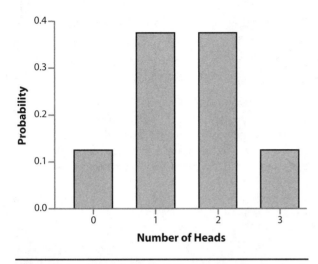

Figure 5.5   Binomial probability distribution for $k = 3$ and $p = 0.5$

Notice that this graph and the preceding one are symmetric; the probabilities for values of x (numbers of heads) are the same on the two sides. This symmetry is because for this case $p = q$. When $p \neq q$, the graph will not be symmetric (dice example below).

You may recall from a course in algebra that expanding binomial expressions is not a lot of fun, especially when k becomes large. Fortunately there are better ways. We can use the binomial formula (below) to calculate exact probabilities. When k gets really large (say > 30), we can also use the normal approximation for the binomial (section 8.5). For our purposes here, we will use the binomial formula. The probability of the occurrence of x events of interest (heads, for example), symbolized by $p(x)$, in k trials (coins tossed, for example), is given by equation 5.9 (also called the **binomial formula**):

$$p(x) = \frac{k!}{x!(k-x)!} p^x q^{(k-x)} \qquad (5.9)$$

where k is the number of trials (coins tossed), x is the number of occurrences of the event of interest whose probability we wish to predict (the number of heads), and p is the probability associated with the occurrence of x (0.5 in this case). Equation 5.9 is derived from two of the rules introduced above—the multiplication rule (second part of the formula) and the addition rule (first part). You should recognize the first part of the formula (also called the "**binomial coefficient**") as the number of combinations of k objects taken x at a time, introduced ear-

lier. Some examples will help to illustrate how equation 5.9 is used.

Example 5.1
Calculating a Binomial Probability: Coin Toss

Suppose we wish to know the probability of obtaining 1 head in the toss of 3 coins (either 3 coins tossed simultaneously or 1 coin tossed 3 times in succession—in other words, in 3 trials). In this case $x = 1$ (the number of heads in which we have an interest), $k = 3$ (the number of trials), $p = 0.5$ (the probability associated with event x), and $q = 1 - p = 0.5$.

Substituting these values in equation 5.9 gives:

$$p(x = 1) = \frac{3!}{1!(3-1)!}(0.5^1)(0.5^2) = 0.375$$

To verify that you understand the use of equation 5.9, find $p(x)$ for the other 3 possible outcomes in this situation. Based on your calculations, you should find that the probability of 0 heads and 3 tails is 0.125, of 1 head and 2 tails is 0.375, of 2 heads and 1 tail is 0.375, and of 3 heads and 0 tails is 0.125. Refer again to figure 5.5.

Example 5.2
Calculating a Binomial Probability: Rolling the Dice

So far we have considered only situations in which p and q are equal (both 0.5). This is not always the case. Consider the probability distribution of obtaining x number of ones in the toss of 2 dice. We will consider a "1" to be the event of interest and "not 1" as the other event. The probability of obtaining any particular number on the roll of a single die is 1/6 (0.1667). Thus, p (the probability of a 1) is 0.1667, and q (the probability of obtaining any result except 1) is 1 - 0.1667, or 0.8333. Since we are tossing 2 dice in this experiment, $k = 2$. Let's again calculate one of the binomial probabilities, say for exactly 1 "one." Substituting these values in equation 5.2 gives this result:

$$p(x = 1) = \frac{2!}{1!(2-1)}(0.1667^1)(0.8333^1) = 0.2778$$

The binomial distribution, showing the probabilities for all possibilities, is shown below. The sum of all these probabilities should equal 1, as it does. Calculate a couple of these probabilities to verify you are using the formula correctly.

Number of "Ones" (x)	Number of "Not Ones"	$p(x)$
0	2	0.6944
1	1	0.2778
2	1	0.0278
	Sum	1.000

Thus, the probability of obtaining 2 ones in a toss of 2 dice is only 0.0278 (or approximately 3 times in 100 tosses). This probability distribution is shown as a bar graph in figure 5.6. Again, notice the lack of symmetry in this probability distribution.

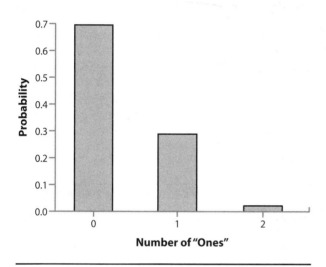

Figure 5.6 Binomial probability distribution for $k = 2$ and $p = 0.16667$

5.4.1 Cumulative Probability

Sometimes we are interested in the chance of obtaining the sum of all the probabilities. (This is done the same way as keeping a running total of points in a game.) In the case of probabilities, if we knew the risk of a genetic disease in any given child and we had a family of x children, we might want to know the chance of one or more affected children. We could compute each binomial probability and add them up.

As a simple example, consider the dice from above and determine the chance of obtaining one or more "ones." For a random variable X and real number x, the cumulative probability is represented as $p(X \leq x)$. Notice that the cumulative probabilities are determined by taking a running total from the $p(x)$ column.

x	p(x)	p(x ≥ X)
0	0.6944	0.6944
1	0.2778	0.9722
2	0.0278	1.0000
Total	1.0000	

5.4.2 A Note about Working Exercises and Using Computer Statistical Packages

If you have done many of the practice exercises, you may find the calculations a bit tedious. Always do enough exercises that you are confident in knowing what you are trying to accomplish and the methods you need to follow to solve a particular problem type. Once you know how to do problems by hand, we encourage you to then explore available computer software. *Minitab* can easily generate probability distributions (both individual and cumulative).[5] We encourage you to run some of the same problems both by hand and with the computer to see that you get the same answer. For the above dice problem, *Minitab* output for the individual probabilities and cumulative probabilities are given below. Notice the difference between the two columns.

Probability Density Function	Cumulative Distribution Function
Binomial with n = 2 and p = 0.166667	Binomial with n = 2 and p = 0.166667
x P(X = x) 0 0.694444 1 0.277778 2 0.027778	x P(X <= x) 0 0.69444 1 0.97222 2 1.00000

5.4.3 Expected Frequencies

Expectations are frequently used in decision theory, which guides management decisions in business, military planning, and sports. (Keep in mind that, because of chance, these decisions are not always right!) In biology, we may be interested in knowing if the number of individuals follows some particular probability distribution. Do the observed numbers from a cross in genetics follow the predicted 9:3:3:1 ratios? Is the distribution of individuals random? Such questions will be addressed more formally with hypothesis testing later. So, for now we'll compare observations to predictions by eye—do they look close or very different?

To compare expected with observed frequencies, we must calculate the expected frequencies from a particular probability distribution. If we have n trials and each value of x has associated probability $p(x)$, then the expected value of x is determined by:

$$E(x) = n * p(x) \qquad (5.10)$$

Below we use expected values of x from two different probability distributions, first from our familiar Binomial distribution and later from the Poisson distribution (section 5.5).

5.4.4 Comparison of Predictions from Binomial Distribution to Observations

Of what practical use is the binomial distribution to us as biologists? Perhaps some examples will serve to illustrate some of the many ways we use this particular probability distribution.

Example 5.3
A Biological Example of a Binomial Variable

Bush hogs always have litters of 6 hoglets. We will assume, based on our knowledge of chromosomal sex determination, that the probability that any hoglet in a litter will be a male is 0.5 and the probability that any hoglet will be a female is also 0.5. We wish to know how many litters in 100 litters will consist of 5 or more males, if our assumption about the equal probability of an individual hoglet being a male or a female is correct.

Note that this problem is exactly equivalent to asking how often we would obtain 5 or more heads in the toss of 6 coins. First we solve equation 5.2 for $x = 5$, which is:

$$p(5) = \frac{6!}{5!1!}(0.5^5)(0.5^1) = 0.0937$$

Thus, the probability associated with exactly 5 males in a litter of 6 is 0.0937. However, our question asked for the probability associated with 5 *or more* males in a litter of 6. So we must also determine the probability associated with 6 males in a litter, which is:

$$p(6) = \frac{6!}{6!0!}(0.5^6)(0.5^0) = 0.0156$$

The probability associated with exactly 6 males in a litter of 6 is 0.0156. The probability associated with either 5 *or* 6 males in a litter of 6 is $p(5) + p(6)$, which is

$$0.0937 + 0.0156 = 0.1093$$

In 100 litters we would therefore expect to find 100 × 0.1093 or approximately 11 litters consisting of 5 or more male hoglets.

In the hoglet example, n was the total number of litters (100).

Let us ask the same question in a slightly different way: In 100 litters, how many of each of the possible combinations of males and females would we expect to find (how many with 6 males and no females, with 5 males and 1 female, with 4 males and 2 females, and so on)?

For this solution we would simply solve equation 5.9 for each value of x from 0 to 6. We would

obtain the results shown in table 5.1. Notice that the sum of the $p(x)$'s is close to 1. (The slight discrepancy is from a rounding error.)

What use might we wish to make of such a probability distribution, which is based on theoretical assumptions rather than on real-world observations? Consider example 5.3. Suppose we suspect that the gender of bush hoglets is not a matter of random chance but rather that some females tend to produce an inordinately large number of offspring of one sex or the other (a phenomenon that has actually been observed in a number of species). Suppose also that we actually surveyed 100 litters of bush hogs, each consisting of 6 hoglets, and found that 20 or so contained 5 or 6 male hoglets and a similar number contained 5 or 6 female hoglets. This is about twice as many as our binomial probability distribution predicts, and therefore, we would have reason to suspect that sex determination in this species is not a random event. In other words, some factor or factors other than the chance combination of X and Y chromosomes is involved.

Table 5.1 Binomial Distribution for Example 5.3

Number of Males (x)	Number of Females (k – x)	p(x)	Expected Frequency p(x) × 100
0	6	0.0156	1.56
1	5	0.0937	9.37
2	4	0.2344	23.44
3	3	0.3125	31.25
4	2	0.2344	23.44
5	1	0.0937	9.37
6	0	0.0156	1.56
	Sum	0.9999	99.99

Note carefully that without the prediction based on the binomial probability distribution, we would not know if the number of predominately single-sex litters that we observed was too large to be produced by random combinations of X and Y chromosomes. In other words, we cannot conclude that a result is *not* due to chance alone unless we know what result is expected if it is due to chance. This concept is the basis for much of statistical hypothesis testing, the subject of much of this book.

5.5 The Poisson Distribution

Another discrete probability distribution often useful to biologists is the **Poisson distribution**. Generally we use the Poisson distribution to predict probabilities of the occurrences of "rare" events (when such occurrences are known to be independent of one another) or to determine if the occurrences of such events are independent of one another. The event of interest must be rare, which is to say that the number of occurrences in any sampling unit must be small *relative to the number of times that it could conceivably occur*. In theory, there is no upper limit on the possible number of occurrences in a single sampling unit. In reality, there must be limits. For example, when trees are at low density (say one or two per hectare), it's possible for them to follow a Poisson distribution. However, when the trees are so abundant that they are bumping into one another, the trees cannot be random in their distribution.

For an event to follow a Poisson distribution, occurrences of the event must be independent of previous occurrences of the event in the sampling unit. We can compare an observed frequency distribution with frequencies produced by a Poisson distribution to see if occurrences are independent. These occurrences may be in space (e.g., locations of plants) or in time (e.g., when a bird sings).

Example 5.4
A Poisson Variable: Spatial Distribution of Plants

In the maple seedling example in chapter 3 (reproduced below as table 5.2), most of the quadrats sampled had no seedlings, several had 1, a few had 2, and so on. We would like to know if the occurrences of maple seedlings are independent events. If so, their frequency of occurrence in the quadrats should follow a Poisson distribution. If they do not follow a Poisson distribution, we conclude that the events are not independent and that the occurrence of a maple seedling in a quadrat in some way influences the occurrences of other maple seedlings in that quadrat.

Table 5.2   Maple Seedlings per Quadrat

Number of Plants/Quadrat	Frequency
0	35
1	28
2	15
3	10
4	7
5	5

---

alter the possible occurrence of other seedlings in a quadrat, the frequency of seedlings per quadrat should follow a Poisson distribution. If the observed frequency distribution strongly deviates from the Poisson distribution, then we have reason to suspect that the occurrences of maple seedlings are not independent events, which is to say that the distribution of maple seedlings is not random, and that some factor or factors in the environment other than chance influences the distribution.

What the Poisson distribution predicts is how many units (quadrats, in this case) are expected to have no occurrences of the event (maple seedlings), how many are expected to have 1, how many are expected to have 2, and so on, if the events are rare and independent. These probabilities are given by the expression

$$p(x) = \frac{\mu^x e^{-\mu}}{x!} \tag{5.11}$$

where μ is the population mean occurrence of the event per sampling unit, e is the base of natural logarithms (2.7183), and $p(x)$ is the probability of x events in a unit. Since we usually have no knowledge of the population mean, it is estimated by the sample mean, and $p(x)$ is computed by equation 5.12

$$p(x) = \frac{\bar{x}^x e^{-\bar{x}}}{x!} \tag{5.12}$$

for probabilities, $p(x)$, of 0, 1, 2, 3, 4, and so on, occurrences per unit. In the maple seedling example, there were 100 quadrats (units) and a total of 141 seedlings (events) in these 100 quadrats. The mean number of occurrences ($\bar{x}$) was therefore 141/100 or 1.41 plants per quadrat. Solving equation 5.12 for relative expected occurrences per unit (seedlings per quadrat) of 0, 1, 2, 3, 4, 5, and 6, gives the results shown in table 5.3 under the column designated "$p(x)$." For example, one of these probabilities (1 seedling/quadrat) was determined as follows:

$$p(x = 1) = \frac{(1.41^1)(e^{-1.41})}{1!} = 0.344$$

As indicated earlier, probability distributions are readily created from computer software. From *Minitab*, the Poisson probabilities for this plant ecology example are as follows, confirming our results above.

Since the number of seedlings that could conceivably occur in a quadrat is quite large, we can consider the observed frequencies to be rare events. If the presence of a seedling in a quadrat does not

Probability Density Function	Cumulative Distribution Function
Poisson with mean = 1.41	Poisson with mean = 1.41

x	P(X = x)	x	P(X <= x)
0	0.244143	0	0.24414
1	0.344242	1	0.58839
2	0.242691	2	0.83108
3	0.114065	3	0.94514
4	0.040208	4	0.98535
5	0.011339	5	0.99669
6	0.002665	6	0.99935
7	0.000537	7	0.99989
8	0.000095	8	0.99998
9	0.000015	9	1.00000
10	0.000002	10	1.00000
11	0.000000	11	1.00000

Multiplying the values of $p(x)$ by the number of sampled units (100 in this case) gives the expected frequencies (from formula 5.10). The results are shown in table 5.3 below. The last column in the table gives the values that were actually observed (from table 5.2).

Table 5.3 **Expected and Observed Frequencies of Maple Seedlings per Quadrat ($\bar{x}$ = 1.41, n = 100)**

Number/ Quadrat (x)	p(x)	Cumulative	Expected	Observed
0	0.244	0.244	24.4	35
1	0.344	0.588	34.4	28
2	0.243	0.831	24.3	15
3	0.114	0.945	11.4	10
4	0.04	0.985	4	7
5	0.011	0.997	1.1	5
6	0.003	0.999	0.3	0

Figure 5.7 is a graphic representation of the expected distribution and the actual observed distribution from table 5.3. The expected distribution, based on the Poisson probability distribution, is the distribution we would expect if the maple seedlings were distributed randomly within the sampled habitat.

A frequent use of the Poisson distribution is testing whether particular events occur independently of one another, and therefore are distributed randomly in space or time. When the events are not random, they may either be "clumped" (also called overdispersed) or "repulsed" (underdispersed) with respect to each other. Notice in figure 5.7 that the expected (Poisson) distribution and the actual observed distribution do not match very well. In this

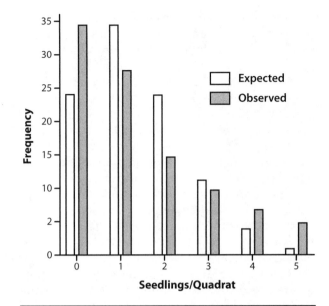

Figure 5.7 Comparison of observed frequencies of maple seedlings per quadrat (from table 5.2) with frequencies expected from a Poisson probability distribution for mean = 1.41.

particular example, the observed distribution has more quadrats than expected with 0 seedlings and more than expected with large numbers of seedlings, which appears like a clumped distribution. We might suspect, therefore, that the occurrences of maple seedlings are not independent of one another. In chapter 7 we will discuss how a goodness of fit test can be applied to determine if this difference is likely real or if it is simply the result of chance.

The maple seedlings example illustrates one way in which the Poisson distribution is commonly used in ecology. But this distribution also finds use in many other situations. Example 5.5 illustrates how the Poisson distribution might be applied to a microbial genetics problem.

Example 5.5
A Poisson Variable: Viruses Infecting Bacteria

Bacterial viruses infect bacterial cells by first adsorbing onto the bacterial cell wall. The number of phage particles that adsorb to any one bacterial cell is a Poisson process. In a certain experiment, 2.5×10^6 phage particles were mixed with 10^6 bacterial cells. Thus, the mean number of phage particles per bacterial cell is 2.5. We wish to know what proportion of the bacterial culture would be expected to have no phage particles adsorbed to them.

The mean number of phage particles per bacterial cell is 2.5. Substituting this value in equation 5.12 and solving for $p(0)$ gives:

$$p(0) = \frac{2.5^0 \times e^{-2.5}}{0!} = 0.0821$$

Thus, the probability that any randomly selected cell would have no phage particles adsorbed to it is 0.0821. This is also the proportion of the cells in the bacterial culture that would be expected to be free of phage particles. The expected number of bacterial cells with no phage particles should therefore be $0.0821 \times 10^6 = 8.21 \times 10^4$.

A question like this can be asked in a different way. Suppose we wish to add a sufficient number of phage particles to a bacterial culture to ensure that no more than 1% (0.01) of the bacterial cells remain uninfected. How many phage particles per bacterial cell would we need to add? Could you suggest how this might be calculated?

Key Terms

binomial distribution
classical probability
combinations
conditional probability
cumulative probability distribution
empirical probability
expected frequency
independent
mutually exclusive
permutations
Poisson distribution
probability
probability distribution

Exercises

SIMPLE PROBABILITY PROBLEMS

5.1 A cage contains 12 brown mice, 8 white mice, and 10 tan mice. A single mouse is chosen at random.

   a. What is the probability of choosing a brown mouse? a white mouse? a tan mouse?

   b. What is the probability of choosing a brown mouse or a tan mouse?

5.2 A jar contains three different colored seeds, 23 brown seeds, 14 green seeds, and 5 yellow seeds. A single seed is chosen at random to be planted.

   a. What is the probability of choosing a brown seed? a green seed? a yellow seed?

   b. What is the probability of choosing a green seed or a yellow seed?

5.3 In the toss of 2 coins, what is the probability of obtaining 2 heads?

5.4 In the toss of 2 coins, what is the probability of obtaining 2 tails?

5.5 In the toss of 2 coins, what is the probability of obtaining 1 head and 1 tail?

5.6 In the toss of 3 dice, what is the probability of obtaining 2 twos and 1 four?

CONDITIONAL PROBABILITY

5.7 For the situation in exercise 5.1, if you remove 2 mice, what is the probability that both will be white? (Note: sampling without replacement)

5.8 For the situation in exercise 5.2, if you remove 3 seeds, what is the probability all three will be brown? (Note: sampling without replacement)

5.9 Suppose you have 24 unlabeled blood samples and your job is to label each tube with the correct blood type. There are 6 of each blood type (O, A, B, AB). If you remove two samples and test them, what is the chance that the first sample is blood type A and the second is blood type O?

5.10 A deck of cards contains 26 black cards and 26 red cards. If you are dealt two cards, what is the probability one card is black and one is red?

THE BINOMIAL COEFFICIENT

5.11 You go into an ice cream store and decide you want three different scoops of ice cream. There are 21 different flavors of ice cream to choose from. How many combinations of three flavors are possible?

5.12 If you toss 10 coins, how many different combinations could possibly give you 6 heads?

THE BINOMIAL DISTRIBUTION

5.13 In the toss of 6 coins, what is the probability of obtaining exactly 2 heads (and thus 4 tails)?

5.14 In the toss of 6 coins, what is the probability of obtaining exactly 3 heads?

5.15 In the toss of 5 dice, what is the probability of obtaining exactly 5 ones?

5.16 In the toss of 5 dice, what is the probability of obtaining exactly 3 twos?

5.17 In the toss of 5 dice, what is the probability of obtaining 2 or fewer ones?

5.18 In the toss of 5 dice, what is the probability of obtaining at least 1 six?

5.19 Assume that the probability that any child born in a population will be a boy is 0.5 (and thus the probability that the child will be a girl is also 0.5). For a randomly selected family of 6 children, what is the probability that there will be exactly 3 boys?

5.20 For the situation in exercise 5.18, what is the probability that there will be 2 or fewer girls?

5.21 In couples where each person is heterozygous for the sickle-cell gene, there is a probability of 0.25 that any child of the couple will actually have the disease, and a probability of 0.75 that any child will not have the disease. For a population of families of 6 children in which both parents are heterozygotes, find the probability that in a randomly selected family:

a. none of the 6 children will have the disease?

b. one or more of the children will have the disease?

5.22 Suppose it is known that in a certain population 10% of the population is color blind. If a random sample of 25 people is drawn, find the probability that:

a. none will be color blind

b. one or more will be color blind

5.23 Construct a table and a bar graph of the probabilities of all possible outcomes of the situation in exercise 5.21 (no children have the disease, 1 child has the disease, 2 children have the disease, and so on).

5.24 Construct a table and a bar graph of the probabilities of all possible outcomes of the situation in 5.19 (0 boys, 1 boy, 2 boys, and so on).

5.25 Autism Spectrum Disorders (ASD) are a group of developmental disorders that affect a person's ability to communicate and behave socially. The probability of a child in the United States having ASD is about 0.0091. Suppose families consist of 3 children. Determine the binomial distribution for this example by hand and using a statistical package. Confirm that your answers are equivalent and then compute the expected frequency of families (out of 1000) in which there are no children with the disorder.

5.26 For the situation in exercise 5.25, compute the expected frequency of families (out of 1000) in which there are exactly 1 out of 3 children with the disorder.

THE POISSON DISTRIBUTION

5.27 Using the ant lion data from exercise 2.3 (see chapter 2), compute the Poisson probabilities for $x = 0$ to $x = 8$ for a population whose mean is the same as this sample mean. Determine these probabilities by hand and with a statistical package. Confirm that your answers are the same.

5.28 Using the results from exercise 5.27:

a. Calculate the expected frequencies and in a table compare them to the observed frequencies (from exercise 2.3).

b. Construct a bar graph of the expected frequencies and observed frequencies

c. Do the observed frequencies appear to match the expected frequencies? What does this suggest about the distribution of ant lion pits (random or not)?

5.29 In a culture containing 3×10^7 bacterial cells and 5×10^7 bacteriophage, what proportion of the bacterial cells would be expected to have:

a. no phage particles adsorbed to their cell walls?

b. 1 or more phage particles adsorbed to their cell walls?

5.30 In a culture containing 5×10^8 bacterial cells and 3×10^8 bacteriophage, what proportion of the bacterial cells would have 3 or more phage particles adsorbed to their cell walls?

Notes

[1] For clarity, we have chosen to use an intuitive description of probability. The meaning of the word will become clearer as the chapter develops. For a more precise definition of probability, see Wikipedia.

[2] The word "die" is the singular of "dice" (one die, two dice).

[3] This basic rule assumes that all events are equally likely. Here we ignore those sampling schemes where the possible outcomes are not equally likely.

[4] For factorials, $n! = n \times (n - 1) \times (n - 2) \ldots \times 1$ (e.g., $3! = 3 \times 2 \times 1 = 6$) and $0! = 1$.

[5] For those of you using *Minitab*, type in the possible values of x in one column (0, 1, 2, ...), then choose menu "Calc" and select "Probability distributions."

chapter six

Statistical Inference and Hypothesis Testing

As we saw previously, statistics are widely used in science (and everyday life) to describe the meaning from a large mass of numerical data. Sometimes we can stop right here. For instance, if we study a small population from which we can collect every individual measurement, then we can directly calculate the parameter of interest. Sports enthusiasts may keep track of batting averages for their favorite baseball players. Fisheries biologists may need to keep track of fungus infections in their hatchery tanks. Draining the tank and counting the number of infected and non-infected fish allows them to determine the % infected. Educators may want to know the average score on an exam for their classes each year, and can simply calculate each mean directly. In each case, since our population was collected in its entirety, we have directly determined the value for the parameter of interest.

On the other hand, if the population is much larger we must instead take a random sample, from which we can make inferences about the entire population. Suppose an ecologist wants to know if the prevalence of chytrid fungus infection in frogs from wetlands in the Rocky Mountains is increasing over time. There is no way that they could capture every frog. Instead, each year, they must collect a proper sample, classify the infection status of each frog, and make some broader inference about the statistical population from that sample. (Refer back to chapter 1 on further examples and considerations for taking random samples.) Understanding the underlying probability distribution for such a characteristic guides our ability to make these inferences. The purpose of this chapter is to introduce the concept of statistical inference, an idea that we will see developed in multiple directions in later chapters.

6.1 Statistical Inference

Statistical Inference refers to the process of making a conclusion about a population based on a sample taken from that population. There are two broad areas of inference: *estimation* and *hypothesis testing*. **Hypothesis testing** refers to using data to evaluate the validity of a particular idea (a hypothesis). For example, does a particular over-the-counter medication offer a greater relief to the symptoms of hangovers than simply drinking water? Is the growth rate of fish living in polluted habitats slower than those living in pristine habitats? Does dominance rank matter in terms of the number of grandchildren that a bull elephant leaves behind? In each of these examples, we are asking: is there an effect of the treatment or not? Hypothesis testing is discussed in some detail in sections 6.2–6.4 below, with many examples in remaining chapters of this book.

Estimation refers to the interval within which a parameter falls at some specified probability. For instance, consider the chytrid fungus infection example above. Suppose we collect 100 Northern leopard frogs (*Rana pipiens*) from a single large wetland and 15 of these frogs are infected. Our sample estimate of the proportion of frogs with infection is: $P = 15/100 = 0.15$. Based on this sample, can we make a statement about the true population infection rate (p)? Is it 0.15? That is our best guess, but we can do better if we can give a likely range that includes p. Based on a method we'll encounter later (section 8.6), we can determine the confidence interval for p. For our fungus example, the true population proportion (p) is likely in the range 0.089–0.234. To be more precise, the range 0.089–0.234 brackets the true infection rate, p, with 95% probability. This is an important point. The true proportion has some true value; we just don't know exactly what it is. We have a better idea now, however, since this particular confidence interval likely includes that true value.

In chapter 9, we will explore estimation for a measurement variable, generating a confidence interval for its population mean (μ) from a sample in which we have calculated the sample mean and standard deviation. Such information is closely associated with formally testing a hypothesis with t tests.

6.2 Statistical Hypotheses

Before we consider hypothesis testing in detail, let's consider its context in the scientific method. Science begins with observations, which are simply events in the physical universe that we may detect in some way. Measurement of a single individual in a population is an observation and its value designated as x. Observations are important, but they tend to not be very useful in and of themselves. It is when we try to explain the observations that science begins. A tentative explanation of one or more observations is a hypothesis. More exactly, a **hypothesis** is an idea that can be tested by observation and experiment. A good hypothesis allows us to generate a specific prediction that can be compared with new data (fig. 6.1).

In hypothesis testing, we attempt to reduce the question at hand to two choices, the **null hypothesis**, symbolized by H_0 ("H naught"), and the **alternative hypothesis**, symbolized as H_a. Only one of these statements is directly tested (the H_0), and the parameter of interest is specified. (A parameter refers to such things as the true population proportion or the true population mean.) An understanding of the nature of the null hypothesis is critical to the understanding of the nature of hypothesis testing. The null hypothesis is the default, the "straw man," which we attempt to disprove. Depending on the question of interest, the null hypothesis can represent no effect (such as from an experimental treatment) or the expectation from some specific probability model. The alternative hypothesis contains all the other possibilities. Table 6.1 illustrates some examples.

Table 6.1   Some Examples of Null and Alternative Hypotheses

Null hypothesis (H_0)	Alternative hypothesis (H_a)
The medication has no effect on heart rate.	The medication affects heart rates.
Body temperature is normal (= 98.6°F)	Body temperature is not normal ($\neq$ 98.6°F)
Body temperature is not elevated ($\leq$ 98.6°F)	Body temperature is elevated (> 98.6°F)
The sex ratio = 1:1 (50% female, 50% male)	The sex ratio $\neq$ 1:1

Notice that the null hypothesis always contains an equal sign, or at least an implied equal sign. The reason for this is that *a null hypothesis must allow us to generate a single probability distribution*. We generate a probability distribution whenever we run a statistical test. Each statistical test has its own probability distribution associated with, such as a chi-square test (chapter 7) or a t test (chapters 9 and 10), whose distributions are known when the null hypothesis is true. The alternative hypothesis is stated as everything other than the null hypothesis. These two hypotheses, H_0 and H_a, are thus mutually exclusive and exhaustive. In other words, when one

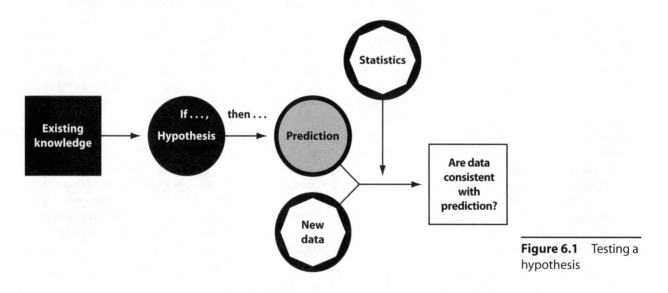

Figure 6.1   Testing a hypothesis

hypothesis is true the other cannot be true, and between the null and alternative hypotheses all possibilities are covered.

Consider a concrete example,[1] based on cases dealing with racial bias in jury selection from the southern United States in the 1960s. In theory, juries are *randomly selected* from a pool of eligible voters. Suppose that 50% of voters in a southern county were African American. The null hypothesis would be that the jury panel was 50% African Americans (i.e., H_0: The percentage of African Americans on the jury = 50%). The alternative hypothesis would be the opposite of the null hypothesis and therefore would state that the panel is composed of some percentage other than 50% African American (i.e., H_a: The percentage of African Americans on the jury ≠ 50%). Actually, we are really only concerned with a negative bias here; so the alternative is really that the panel is < 50% African American. If we then obtain a sample of 40 eligible voters and find only 10 African Americans on the panel, is this sufficient evidence to conclude racial bias? Using the Binomial probability distribution ($n = 40$, $p = 0.5$), the probability of finding 10 (or fewer) African Americans is 0.00111. What this means is that, if the null hypothesis (no bias) were true, the chance of getting such an extreme outcome is highly unlikely! We would conclude that this panel was not drawn at random and has a much lower number of African Americans than expected by chance, and we would reject the null hypothesis.

Science does not proceed by directly proving things. Science instead proceeds by disproving things and, eventually, incorrect theories will be proven to be incorrect. This business of attempting to disprove hypotheses and theories is what we call the scientific method. *If* the hypothesis is correct, *then* this (prediction) will be the outcome of the experiment. If the outcome of the experiment is something other than what the hypothesis (theory) predicts, then we reject the hypothesis and look for a better explanation. The role of statistics is to use quantitative data and determine probabilities to improve our efficiency of weeding out the bad ideas.

Note that we never "accept" the null hypothesis; instead, we only fail to reject it. Why such a negative attitude? Philosophers of science tell us that a false hypothesis may be proven to be false, but a true hypothesis may not be proven to be true.[2] If the hypothesis has general or widespread application to events in the physical universe, we designate it as a **theory**. Like a hypothesis, a true theory cannot be proven to be true, but a false theory can be proven to be false. Lamarck's theory of the inheritance of acquired characteristics was a pretty good theory for its time—it simply failed to stand up to repeated testing (experiment), and so it was abandoned in favor of a better theory.

6.3 Statistical Decisions and Their Potential Errors

The jury pool example above indicated that, if that jury had been assembled at random, then the chance of such an extreme outcome was only 0.00111. This is a good example of a p-value. In any statistical test, the **p-value** is the probability of obtaining a test statistic as extreme (or more extreme) if the null hypothesis was true. When conducting a hypothesis test, we should decide ahead of time the strength of evidence needed to reject the null hypothesis. We choose a probability called alpha (α) and if our p-value is less than alpha, we will reject H_0. (The concept of looking up "critical values" for the various statistical procedures described in this book is exactly the same thing.) Most investigators set $\alpha = 0.05$. There are special circumstances when investigators choose a higher or lower probability (say 0.10 or 0.01), but most often 0.05 is used and we will follow that proportion in this book.

In any statistical procedure, it is important to always report the p-value. Let's suppose that we ran a test procedure and got a p-value of $p = 0.02$. Is H_0 true or false? If we had chosen $\alpha = 0.05$, we would reject H_0. But suppose we had been more conservative and had chosen $\alpha = 0.01$. In that case, we would have failed to reject H_0. The fact that it is possible to reach 2 contradictory conclusions from the same data, depending on where we set alpha, should convince you that there is always an element of subjectivity involved in statistical inference. Reporting the p-value allows others to make their own conclusions based on the strength of the data presented.

The example here brings up an important point. The null hypothesis is in fact either true or false; we are just not completely sure which choice is correct. And, in any statistical test, there are two decisions that we can make based on what the data tell us: either reject H_0 or fail to reject H_0. Thus there are four possibilities, in two we make the correct decision and in the other two we make an incorrect decision. The two types of errors are called the type I error and the type II error. Table 6.2, sometimes called a "truth table," illustrates the relationship between the null hypothesis and the type I and type II errors.

Rejecting the null hypothesis when it is in fact true is called a **type I error**. The risk that such an er-

Table 6.2 The Possible Consequences of a Statistical Decision

		DECISION FROM THE TEST	
		Fail to Reject H_0	Reject H_0
REALITY	H_0 True	Correct	Type I Error
	H_0 False	Type II Error	Correct

ror will occur is called the **alpha (α)**, and as stated before, we set this value ahead of time (usually to $\alpha = 0.05$) by choosing the critical value for the test. A **type II error** is failing to reject a null hypothesis when in fact it is false. The probability of making a type II error is called **beta (β)**. The inverse of beta $(1 - \beta)$ is called **power**. Power is often not determined in statistical tests. However, more scientists are seeing the benefit in estimating its value. We will explore this topic in more detail in chapter 10 (Power of the Test).

Let's consider a simple example of a type II error. A company discharges toxic chemicals in a stream and would like to avoid costly plant upgrades that would be required if they were to stop the discharges. A lax regulatory agency leaves the testing up to the company and the company collects stream samples from 4 locations above the discharge pipe and 4 locations below the discharge. The company environmental scientist examines the diversity of stream macroinvertebrates, a characteristic that can be quite variable from place to place. The scientist finds that the mean diversity is 6.2 above the plant and 3.6 below the plant, but finds no significant difference between in the mean diversity between upstream and downstream sites. The company reports that their pollution level is safe for the "stream health." In reality, there may very well be a large effect of this company on the stream, but, given their sampling design, the power of their test is quite low and they likely made a type II error.

Important points about alpha (α) and beta (β) that you should keep in mind are as follows:

1. Alpha is fixed by the investigator at a certain level, usually 0.05 (or sometimes 0.10 or 0.01). Therefore, its value is known. If the probability associated with our test (p-value) is equal to or smaller than alpha, the null hypothesis is rejected.

2. The value of beta, on the other hand, cannot be known exactly. However, for any given test, beta decreases (power increases) as sample size increases. Thus, large samples result in a smaller beta risk (and hence higher power) than do smaller samples.

3. Occasionally, when more than one statistical procedure is appropriate for the data, some tests may be more powerful than others. When one has a choice of two or more tests for testing the same null hypothesis, the most powerful test should always be used.

4. Designating a lower level for alpha decreases the risk of a type I error (by definition), but at the same time it increases the risk of a type II error (beta risk). One reason alpha is so frequently set at 0.05 is that this level represents a fairly reasonable compromise between the type I error and the type II error.

You have likely heard or read of results of a statistical test being referred to as being "significant" or perhaps even "highly significant." The expression "**significant**" usually means that H_0 may be rejected at an alpha level of 0.05 (i.e., when the p-value associated with the test statistic is 0.05 or less). "Highly significant" usually means that H_0 may be rejected when alpha is set at 0.01 (the p-value is 0.01 or less).

This particular terminology is most unfortunate, since most people who use the English language understand (correctly) that "significant" means approximately the same thing as "important," "meaningful," or "of some consequence." Thus, when some scientist informs you that his or her results are "highly significant," try not to be overly impressed! This simply means that he or she has very likely reached the proper conclusion about a question that might be important or that might be totally trivial. A better expression to use is "**statistically discernible**."

Fortunately, use of the term "significant," and even the use of predetermined alpha levels, seems to be losing popularity in favor of simply expressing the p-value associated with the test statistic. Thus, in our jury example, we could instead simply report that p-value was 0.00111. (This result is clearly statistically discernible.)

6.4 Application: Steps in Testing a Statistical Hypothesis

Hypothesis testing follows a logical sequence of steps. In solving the exercises at the end of subsequent chapters, you should follow this sequence:

1. State, very clearly, the question you are attempting to answer.

2. What sampling distribution describes a sample of this kind, and what is the appropriate statistical test?

3. State the null hypothesis (H_0) and the alternative hypothesis (H_a).

4. Determine the level of alpha at or below which you will reject the null hypothesis, and locate the critical value from the appropriate table. (This is sometimes referred to as the decision rule.)

5. Using the data at hand, calculate the test statistic.

6. Using the appropriate table (or computer software), determine the p-value.

7. Make your decision about H_0: if the probability of obtaining this calculated value is equal to or smaller than the pre-selected value of alpha, reject the null hypothesis and accept the alternative hypothesis.

8. Interpret your decision in light of the original question (#1 above).

Along with doing a hypothesis test, one should always provide some additional supporting evidence, such as a graph or table of descriptive statistics (chapters 1, 3, and 4). This allows you (and a reader) to visualize whether or not your answer makes sense. If your graph shows that a mean is clearly different from a hypothesized value and your test fails to reject H_0, then you may have made an error somewhere.

6.5 Application of Statistics to Some Common Questions in Biology

Hypothesis testing is widely used in Biology. We already saw a variety of data types in the graphics of chapter 1. Each of these data types has a particular statistical method appropriate to its analysis. Cate-

gorical variables and their frequencies are analyzed with chi-square tests (chapter 7). Measurement variables compared between groups (categories) may be analyzed with t tests (chapter 10) or Analysis of Variance (chapters 11–12). And comparing two measurement variables may be explored with correlation and regression (chapters 13–14). Much of the book will describe these methods in detail. Each of these methods rests on a firm foundation of understanding hypothesis testing.

Key Terms

alpha (α)
alternative hypothesis (H_a)
beta (β)
critical value
hypothesis testing
null hypothesis (H_0)
p-value
power (= power of the test)
significant
statistically discernible
statistical inference
test statistic
type I error
type II error

Notes

[1] This example is modified from a problem presented in the delightful little book by Larry Gonick and Woollcott Smith called *The Cartoon Guide to Statistics* (New York: HarperPerennial, 1993).

[2] F. J. Ayala and B. Black, "Science and the Courts," *American Scientist* 81 (1993): 230–239. A fascinating discussion of the nature of science and hypothesis testing in an article about the role of science in deciding the sort of evidence admissible in the courtroom.

chapter seven

Testing Hypotheses about Frequencies

In many types of biological and medical research, we are interested in the frequencies with which events or objects occur, as they are classified into different attributes. In genetics we commonly ask: do the frequencies of observed phenotypes conform to an expected distribution from a particular mode of inheritance? In medicine, epidemiologists and other public health workers frequently ask about the distribution of diseases in populations. For example, does the incidence of hepatitis A differ between people in Chicago and St. Louis? In this chapter, we will consider statistical tests used with frequency data. We are generally interested in two types of problems: testing the goodness of fit of observed frequencies to a particular distribution and testing the association of frequencies between different groups. We will use the most common procedure, called the chi-square test, for both types of problems. We will also introduce the Fisher exact test for the situation where one of the assumptions of the test of association is not met.

7.1 The Chi-Square Goodness of Fit Test

We are often interested in determining if a frequency distribution of a sample from a population fits or does not fit some expected theoretical distribution, such as a Poisson, binomial, or normal distribution—or for that matter, any sort of distribution we care to specify. Several tests, called **goodness of fit tests**, are designed for this purpose.

The null hypothesis for a goodness of fit test is that an observed frequency distribution is not different from some specified distribution and that any departure of the frequency distribution of the sample from the specified distribution is therefore due to chance alone.

You are probably already acquainted with the chi-square goodness of fit test from genetics. The chi-square goodness of fit test determines how well a set of observed frequencies in two or more mutually exclusive categories fit (or do not fit) some specified (expected) distribution. It is a nonparametric test and requires only nominal measurement plus their frequencies. Other variable types can also be used after pooling into categories (e.g., frequency distributions of continuous measurements).

Assumptions of the Test

1. Data are frequencies, placed in mutually-exclusive categories.
2. Observations are independent of one another.
3. Each category has large enough expected frequencies. When there are 4 or fewer categories, none of the expected frequencies may be less than 5; when there are 5 or more categories, not more than 20% of the categories have an expected frequency of less than 5 and none have an expected frequency of less than 1.

Example 7.1
A Chi-Square Goodness of Fit Test

Two purple-flowered pea plants, both heterozygous for flower color, were crossed, resulting in 78 purple-flowered offspring and 22 white-flowered offspring. Does this outcome differ from a 3:1 ratio of purple-flowered to white-flowered offspring?

Color is controlled by a single pair of alleles, and purple is dominant over white. Because of this we expect a ratio of 3 purple-flowered offspring to 1 white-flowered offspring, based on a binomial probability (section 5.4). Since this hypothesis generates a single probability distribution, we use this statement as our **null hypothesis** and all other statements as the alternative hypothesis (H_0: 3:1 ratio, H_A: not 3:1 ratio). With 100 total plants, ¾ would be 75, ¼ would be 25. Thus, we would expect 75 purple-flowered plants and 25 white-flowered plants. However, by chance alone we might expect that there would be some discrepancy between the expected results and what we actually obtained.

The question in a goodness of fit test is this: is this observed discrepancy between the observed frequencies and the expected frequencies too large to be attributed to chance alone? In other words, is the difference between the observed result and the expected result significant (statistically discernible)? The chi-square statistic is given by

$$x^2 = \sum \frac{(o-e)^2}{e} \qquad (7.1)$$

The steps in its computation, shown in table 7.1, are as follows.

1. Subtract each expected value (column 2) from each observed value (column 1). The difference $(o-e)$ is shown in column 3.

2. Square the values of $(o-e)$, to obtain the squared differences, $(o-e)^2$, as shown in column 4.

3. Divide each squared difference by the expected value (e) for that category to obtain the value of $(o-e)^2/e$ for each category (column 5).

4. Sum the values obtained in column 5 to obtain chi-square.

Table 7.1   Calculation of the Chi-Square Statistic

(1) observed	(2) expected	(3) $(o-e)$	(4) $(o-e)^2$	(5) $\frac{(o-e)^2}{e}$
78	75	3	9	0.12
22	25	–3	9	0.36

$$x^2 = \sum \frac{(o-e)^2}{e} = 0.48$$

To evaluate the chi-square statistic, we'll need to know the number of degrees of freedom for the test.

The degrees of freedom (df) for a goodness of fit test depends on the number of mutually exclusive groups (classes) and the number of parameters estimated from the data (equation 7.2):

$$\text{df} = \text{\# classes} - 1 - \text{\# fitted parameters} \qquad (7.2)$$

An example of such a parameter would be estimating the true population mean, μ, from the sample mean (see example 7.2 below). Example 7.1 above had no parameters estimated from the data. Thus, for this situation, df = # classes – 1.

We now consult table A.3, "Critical Values of the Chi-Square Distribution." The top row of this table gives probabilities associated with values of chi-square, and the left column gives degrees of freedom. Our calculated value of the chi-square statistic, 0.48 with one degree of freedom, is less than the critical value for alpha 0.05 and one degree of freedom (3.84). Take our analysis one step further and determine the p-value. (Recall that the p-value represents the chance of getting such an outcome solely due to chance.) Refer to table A.3 and examine the critical values at one degree of freedom. Notice that the value of 0.48 would fall between the values 0.46 and 2.71. Read the associated probabilities at the top of the table, 0.5 and 0.1. Thus the p-value is: $0.1 < p < 0.5$. (CAUTION: be careful with the direction of the inequalities.) In other words, if the null hypothesis were true, such an outcome would occur between 10% and 50% of the time we ran such experiments. (Imagine an army of researchers all doing the same experiment.) This outcome is quite likely, certainly well above our standard cutoff at alpha of 5%. Therefore, we *fail to reject* the null hypothesis and conclude that the observed frequency does not differ from the expected frequency, confirming our genetic model.

Finally, let's confirm that the assumptions of the chi-square test were met. (1) Categories are mutually exclusive of one another. Flowers can either be purple or white, not both; so Mendel or whoever collected these data could tally each flower into only one category. (2) Observations are independent. In this case, we must assume that only one flower was collected from each plant. If multiple flowers had been collected from each plant, the flowers would not be independent of one another. (3) Each category has large enough expected frequencies. The expected frequencies of 75 and 25 are each larger than 5.

Now let us look at a somewhat more complex situation.

Example 7.2
A Chi-Square Goodness of Fit Test

Refer to the maple seedlings per quadrat example from section 5.5 (Poisson distribution), which is reproduced here as table 7.2. We would like to know if the seedlings are randomly distributed or non-randomly distributed in the sampled habitat. (Recall that when objects or events are randomly distributed, they tend to follow a Poisson distribution.)

Table 7.2   Expected and Observed Values of Maple Seedlings per Quadrat ($n =$ 100, $\bar{x} = 1.41$). Note that the last three rows are pooled for calculating chi-square. (Data from table 5.3)

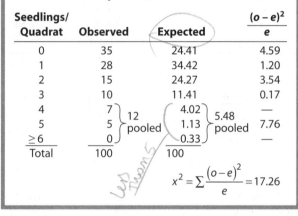

Seedlings/ Quadrat	Observed	Expected	$\frac{(o-e)^2}{e}$
0	35	24.41	4.59
1	28	34.42	1.20
2	15	24.27	3.54
3	10	11.41	0.17
4	7 } 12 pooled	4.02 } 5.48 pooled	—
5	5	1.13	7.76
≥6	0	0.33	—
Total	100	100	

$$x^2 = \sum \frac{(o-e)^2}{e} = 17.26$$

If the 100 maple seedlings in the sampled area were randomly distributed, we would expect, on the basis of the Poisson distribution, to obtain results close to those given in the column headed "Expected." The question is: "How close is close enough?" The expected values are derived from our expectation that the plant locations will follow a Poisson distribution. This is the null hypothesis. If we reject the null hypothesis, we must accept the alternative hypothesis that the seedlings are not randomly distributed. The observed results are shown in the column headed "Observed." Inspection of the data suggest that there are more values of 0 and of 4 or greater than we would expect. To test our null hypothesis, we compute the chi-square statistic as we did before in example 7.1, but with one small difference. Because assumption 3 is violated (low expected values) for the last 3 classes, we should combine the last three values in both the observed and expected columns ("pool the data"). Table 7.2 above gives the results of the chi-square computations for this example. You should do your own calculation to check that the last column and χ^2 are calculated correctly above.

At first glance, there appear to be 4 degrees of freedom in this case (5 groups – 1). However, when it is necessary to estimate one or more parameters in the population of interest, one degree of freedom must be deducted for each parameter estimated (equation 7.2). Recall from section 5.5 that our expected values are based on our sample mean of 1.41 seedlings per quadrat. In other words, we are using an estimate of the population mean (μ) to obtain the expected values. Thus, the degrees of freedom are 5 – 1 – 1 = 3.

Reference to table A.3 gives a critical value of x^2 at alpha 0.05 and 3 degrees of freedom as 7.82. Since our calculated value is larger than this, we reject the null hypothesis. Find the p-value as we did before. (First find the numbers at 3 degrees of freedom that bracket 17.26; then, go to the top of the table and read the probabilities associated with these two numbers.) You should find that the p-value is: $0.0001 < p < 0.001$. State in words what this means. Refer to the interpretation of example 7.1 as need be.

We thus conclude that the observed distribution does not follow a Poisson distribution and hence also conclude that the maple seedlings are not randomly distributed in the sampled habitat. Refer back to the example in section 5.5, so you understand the interpretation of these results.

7.2 The Chi-Square Test for Association

Tests of association are used to determine if two categorical variables are related or associated in some way. (These are also called **heterogeneity chi-square** tests and **chi-square tests of independence**.) Consider the following example.

Example 7.3
A Chi-Square Test for Association

In certain parts of West Africa where malaria is prevalent, there is a mutant form of hemoglobin called sickle-cell hemoglobin or hemoglobin S. Individuals who are homozygous for the hemoglobin S allele develop sickle-cell anemia, an often fatal disease. Individuals who are heterozygous exhibit some mild symptoms of anemia but have an abnormally high resistance to the malaria parasite. Individuals who are homozygous for normal hemoglobin are highly susceptible to malaria. The now infamous 1940s experiment that confirmed this is a good example of the relationship between two nominal variables—genotype and susceptibility to malaria—

(continued)

although the experiment itself would be regarded as inhumane and unethical by present standards. In this experiment, 30 prison inmate volunteers at Stateville Penitentiary near Joliet, Illinois, were selected. Fifteen of these individuals were heterozygous for hemoglobin S, and 15 were homozygous for normal hemoglobin. Otherwise, all 30 were from the same general population and were therefore genetically similar in other respects. All 30 were artificially infected with the malaria parasite. Of the 15 homozygous individuals, 13 contracted the disease, and two did not. Of the 15 heterozygous individuals, one contracted the disease, and the remaining 14 did not.

Such data are customarily displayed in the form of a matrix, sometimes called a **contingency table**, as shown in table 7.3. Notice, in contrast to the observed frequencies in a goodness of fit test shown in different categories as rows, the observed frequencies for a test of association are displayed in two dimensions.

Table 7.3   Association between the Hemoglobin S Allele and Resistance to Malaria

	Contracted Malaria	Did Not Contract Malaria	Total
Heterozygotes	1	14	15
Homozygotes	13	2	15
Total	14	16	30

The question asked in such an experiment is this: is there a relationship between the two variables? Remember, the two variables in this case are both categorical in nature: genotype and susceptibility to malaria. The null hypothesis is that there is no relationship between genotype and susceptibility to malaria (i.e., that the two variables are independent).

Assumptions for the chi-square test of association are similar to those we saw earlier for the goodness of fit test.

Assumptions of the Test

1. Data are frequencies, placed into mutually-exclusive cells.

2. Observations are independent of one another.

3. Not more than 20% of the cells may have an expected value of < 5, and no cell may have an expected value < 1. For a 2 × 2 contingency table, all cells must have an expected value of 5 or greater.

Before we can calculate a chi-square statistic, we need to calculate expected values for each of the

4 categories (2 × 2 = 4). The genotype and malaria results are given again in table 7.4 below, with expected values added. We'll next show how to calculate these expected values.

Table 7.4   Association between the Hemoglobin S Allele and Resistance to Malaria. For the cells, expected values are shown in parentheses; for row and column totals, symbols as given in equation 7.3 are shown in parentheses.

	Contracted Malaria	Did Not Contract Malaria	Total
Heterozygotes	1 (7.00)	14 (8.00)	15 (= R1)
Homozygotes	13 (7.00)	2 (8.00)	15 (= R2)
Totals	14 (= C1)	16 (= C2)	30 (= N)

In a test for association, the expected value for each cell (E_{ij}) is obtained by multiplying its row total (R_i) by its column total (C_j), and then dividing the product by the grand total (N)

$$E_{ij} = \frac{R_i * C_j}{N} \qquad (7.3)$$

The row and column totals are designated accordingly in table 7.4. For instance, the expected frequency of heterozygotes contracting malaria (E_{11}) is:

$$\frac{15 \times 14}{30} = 7.00$$

By the way, equation 7.3 for computing expected frequencies follows the multiplication rule of probability [$P(A*B) = P(A) * P(B)$ (equation 5.4)], which is true when the component probabilities are independent.[1] We next compare the observed result with the result predicted by the null hypothesis, which is that there is no relationship between the heterozygous condition for this gene and resistance to malaria. We calculate a value for chi-square as before: by summing the values $(o - e)^2/e$ for each of the cells. Chi-square for this example is

$$x^2 = \frac{(1-7)^2}{7} + \frac{(14-8)^2}{8} + \frac{(13-7)^2}{7} + \frac{(2-8)^2}{8} = 19.286$$

The degrees of freedom for a chi-square test for association are:

$$df = (rows - 1) \times (columns - 1) \qquad (7.4)$$

In this case $(2 - 1) \times (2 - 1) = 1$. Consulting table A.3, we find that the critical value of chi-square at alpha 0.05 and one degree of freedom is 3.84. Since 19.286

> 3.84, we reject the null hypothesis. Placing our statistic on the table at 1 df, we determine the p-value is $p < 0.0001$. Clearly, this p-value represents strong grounds against the null hypothesis. Interpret the p-value as we have done before. We thus conclude that there is a strong association between heterozygosity for the sickle-cell gene and resistance to malaria. In our initial question, we wanted to determine if the heterozygous genotype conferred some resistance to malaria. Using the chi-square test of association, we determined simply that genotype is associated with resistance. We cannot say which direction the change is, although, looking at the table, the data conform to our initial expectation (heterozygotes confer resistance). Most scientists stop here, simply showing the contingency table, the chi-square statistic, and p-value, and reporting the direction of change.

Chi-square tests are very easy to do with computer programs. A goodness of fit test can be simply done with a spreadsheet. (This is very handy when doing multiple practice exercises; just replace the raw numbers!) *Minitab* does the chi-square test of association and shows all the elements of the analysis, as well as reporting the p-value and showing each term's contribution to the chi-square statistic. The *Minitab* output for the example just conducted follows in table 7.5. Check to confirm that the expected values and chi-square statistic are the same as already computed. Note that the p-value reported (0.000) should be interpreted as $p < 0.001$.

The chi-square test for association need not be limited to 2 × 2 contingency tables, but may be expanded to as many rows and columns as necessary. Each row or each column would represent one mutually exclusive category into which an individual might fall. When there are more than two mutually exclusive categories, the computational procedure is exactly the same as demonstrated above. The only change is in the number of degrees of freedom (equation 7.4).

7.3 The Fisher Exact Probability Test

The Fisher exact probability test, like the chi-square test for association, is used to test for an association between two categorical variables. An important difference is that the requirement that no cells have an expected value of less than 5 is not applicable to the Fisher test, and accordingly, the test is very useful for small samples. The test is limited to 2 × 2 contingency tables.

Assumptions of the Test

1. Data are frequencies, placed into mutually-exclusive cells.
2. Observations are independent of one another.
3. The table has 2 rows and 2 columns.

Example 7.4
The Fisher Exact Probability Test

An experiment was conducted to determine if neonatal garter snakes were more or less likely to exhibit an avoidance response to a threatening stimulus presented from above or from "snakes-eye level." One sample of seven snakes was presented one at a time with an overhead stimulus, and another sample of seven was presented with the same stimulus presented laterally. The investigator recorded whether the snakes responded by attempting to "escape" or failed to respond. (Data from R. Hampton and J. Gillingham)

Table 7.5 Chi-square Test of Association Results from *Minitab*. Hemoglobin-malaria data from Example 7.3.

Chi-Square Test: Malaria, No malaria

```
Expected counts are printed below observed counts
Chi-Square contributions are printed below expected counts

                     No
        Malaria   malaria   Total
   1          1        14      15
          7.00      8.00
          5.143     4.500

   2         13         2      15
          7.00      8.00
          5.143     4.500

Total       14        16      30

Chi-Sq = 19.286, DF = 1, P-Value = 0.000
```

There are 2 independent samples involved here—snakes which were stimulated from above and snakes which were stimulated laterally—and 2 mutually exclusive categories—responded or did not respond. To conduct the Fisher test, data of this type are arranged in a 2 × 2 contingency table, as in tables 7.6 and 7.7 below. Notice that, because of the small sample sizes, the chi-square test of association is inappropriate (assumption 3 above).

Table 7.6 Arrangement of Data for the Fisher Exact Probability Test

	Classification I	Classification 2	Totals
Group I	A	B	A + B
Group II	C	D	C + D
Totals	A + C	B + D	A + B + C + D

For the Fisher exact test, when the null hypothesis of no association is true, the probability of observing this particular distribution is given by:

$$p = \frac{(A + B)\,!\,(C + D)\,!\,(A + C)\,!\,(B + D)\,!}{n!\ A!\ B!\ C!\ D!} \quad (7.5)$$

where $n = A + B + C + D$, and by convention $0! = 1$.

The data for the example are shown in table 7.7.

Table 7.7 Response of Neonatal Garter Snakes to Overhead and Lateral Stimuli

	Responded	No Response	Totals
Overhead Stimulus	6	1	7
Lateral Stimulus	1	6	7
Totals	7	7	14

The probability of this outcome under the null hypothesis is

$$p = \frac{7!\,7!\,7!\,7!}{14!\,7!\,0!\,0!\,7!} = 0.0002914$$

This is the probability of the observed distribution if H_0 were true. However, we are not interested in exactly this outcome, but of this outcome or of any more extreme outcomes with the same marginal totals. There is only one more extreme outcome in this case, which is:

	Responded	No Response	Totals
Overhead Stimulus	7	0	7
Lateral Stimulus	0	7	7
Totals	7	7	14

The probability for this case is

$$p = \frac{7!\,7!\,7!\,7!}{14!\,7!\,0!\,0!\,7!} = 0.0002914$$

Thus, the probability under H_0 of our observed distribution or of one even more extreme is 0.01428 + 0.00029 = 0.01457. Based on this small p-value, which is less than our usual level of α (0.05), we may reject H_0 and conclude that neonatal garter snakes are more responsive to the overhead stimulus than to the eye-level stimulus.

The expected values in this example would be too small for the chi-square test for association. Therefore, the Fisher exact probability test was used. The Fisher test may also be used when the expected values are large enough to permit use of the chi-square test for association. However, the cumbersome calculations of the Fisher exact test make the chi-square test a much-simpler alternative!

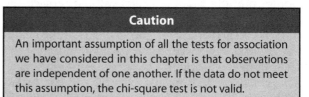

Caution

An important assumption of all the tests for association we have considered in this chapter is that observations are independent of one another. If the data do not meet this assumption, the chi-square test is not valid.

Independent means that any particular event has no influence on the probability of other events occurring. What exactly does this mean? Consider an animal behavior study with cattle. A researcher counted the frequency with which bulls exhibited one of three reproductive behaviors in the presence of either conceptive or non-conceptive cows (fig. 7.1). The reported total number of occurrences (N) was 213. Thus, we would assume that 213 bulls were observed one time, each on approaching a cow in one of the two conditions of the test. However, careful reading of the study indicated that only 14 bulls were used in this particular experiment. Accordingly, each bull was observed an average of over 15 times! This method of using the same subject many times, but treating each event as independent is clearly incorrect. The researchers might have been able to resurrect their results using a different statistical method. However, they did not keep track of which bull did what, so we have no knowledge of how many times each bull was actually observed. Some may have been observed many times (meaning that they would make a large contribution to the results), while others might have been observed only a few times.

This particular example gives us an important lesson. While designing experiments, researchers

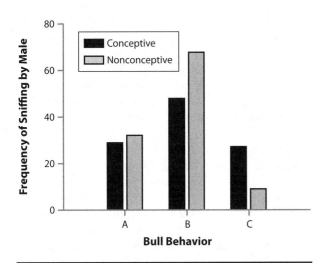

Figure 7.1 Frequency of olfactory behaviors, classified by reproductive stage of female.

do well to consider some basic principles of experimental design (chapter 15) and to seek advice from those knowledgeable about statistics.

Key Terms

chi-square test of association
chi-square test of independence
contingency table
Fisher's exact test
goodness of fit test
heterogeneity chi-square
independence

Exercises

Exercises preceded by an asterisk require a computer for solution. Work some other exercises in each category by hand and also use a computer to confirm you get the same answer. For all hypothesis tests, be sure to specify the null (H_0) and alternative hypotheses, conditions for rejecting the H_0 (at $\alpha = 0.05$), your resulting test statistic, p-value, decision about validity of H_0 (reject or fail to reject), and verbal conclusion.

Chi-Square Goodness of Fit

7.1 One hundred chickadees were given a choice of either striped sunflower seeds or black sunflower seeds. Black seeds were chosen by 75. May we conclude that the population from which this sample was taken has a preference for black sunflower seeds over striped sunflower seeds?

7.2 Suppose that in the situation described in exercise 7.1, 48 chickadees had chosen black seeds, and 52 had chosen striped seeds. May we conclude that there is a preference for striped seeds?

7.3 In a genetic experiment involving flower color in phlox, a ratio of 3 blue-flowered plants to 1 white-flowered plant was expected. The observed results were 35 blue-flowered plants and 14 white-flowered plants. Does the observed ratio differ significantly from the expected ratio?

7.4 Suppose that in exercise 7.3 a ratio of 61 blue-flowered plants to 22 white-flowered plants was observed. Does the observed ratio differ significantly from the expected ratio?

7.5 In 102 tosses of 4 coins, the following results were obtained. (Note that you'll need to use the Binomial distribution [section 5.4] to determine expected frequencies.)

Heads	Tails	Observed Frequency
4	0	8
3	1	23
2	2	40
1	3	27
0	4	4

Does this outcome differ significantly from what would be predicted by chance alone?

7.6 In 120 rolls of 5 dice, the following results were obtained.

Number of Sixes	Observed Frequency
5	6
4	8
3	12
2	20
1	39
0	35

Assuming that the dice are "fair," is it likely that this outcome would occur by chance alone? (Or more to the point, would you care to gamble with the owner of these dice?)

7.7 Refer to exercise 2.3 concerning the distribution of ant lion pits. Using the sample mean as the best estimate of the parametric mean, determine if the distribution of pits follows a Poisson distribution. Interpret your results biologically. (Note: When determining the Poisson distribution carry out x until the cumulative probability equals 1. Use a statistical program to determine the Poisson Distribution.)

7.8 A study in Southern Brazil assessed the abundance of terrestrial flatworms by surveying 100 plots. The following table shows the observed number of flatworms per plot. Using the sample mean as the best estimate of the parametric mean, determine if the distribution of flatworms is random. (Note: When determining the Poisson distribution carry out x until the cumulative probability equals 1. Use a statistical program to determine the Poisson Distribution.)

Number of flatworms/plot	Observed Frequency
0	52
1	33
2	12
3	3

Data adapted from Antunes et al. 2012. *Pedobiologia* 55: 25–31

CHI-SQUARE TEST OF ASSOCIATION

7.9 It is suspected that female water snakes that forage in Lake Michigan migrate to inland ponds in the fall to deliver their young. If this is correct, one might expect that females would be much more likely to migrate at that time than would males. The following data were collected.

	Migrators	Nonmigrators
Females	25	2
Males	4	30

Data from C. Meyers

Is there an association between sex and migration?

7.10 White-throated sparrows occur in 2 distinct color morphs, referred to as brown and white. It was suspected that females select mates of the opposite morph (i.e., white females select brown males and vice versa). This phenomenon is known as negative assortative mating. In 49 mated pairs, the color combinations were as follows.

		Males	
		White	Brown
Females	White	7	23
	Brown	14	5

Data from D. Tuzzalino

Do the results support the assumption that negative assortative mating occurs in this species?

*7.11 Using the data from Digital Appendix 3, determine if male university students are more or less likely to smoke than are female university students (i.e., determine if there is an association between smoking and sex).

*7.12 Using the data from Digital Appendix 3, determine if male university students are more or less likely to do aerobic exercises on a regular basis than are female university students (i.e., determine if there is an association between sex and aerobic exercising).

7.13 An article in the *Springfield News Leader* (11/7/03) summarized results from a British study, which concluded that marijuana (cannabis) relieves symptoms of multiple sclerosis (MS). The study stated that 630 MS patients were included in an experiment in which they were randomly assigned to one of three treatments, with equal numbers in each treatment. The treatments were: cannabis oil, synthetic cannabis (THC), and a placebo control. This was a double blind study (i.e., neither patients nor their doctors know which treatment they received). The following numbers are *percentages* of each group which experienced symptom relief. For each disease symptom, test the null hypothesis that treatment and symptom relief are independent. Do the results support the conclusions in the article?

Treatment	Pain relieved (%)	Stiffness relieved (%)
Cannabis	57	61
THC	50	60
Placebo	37	46

7.14 The frequency of 3 food items (small snails, cladocerans, and mosquito larvae) in the stomach contents of male and female mosquito fish was determined. Do males and females differ with respect to food items utilized?

		Food Item		
		Snail	Cladoceran	Mosquito Larvae
Sex	Males	50	23	15
	Females	10	14	62

7.15 Suppose that the outcome of exercise 7.10 had been as follows.

		Males	
		White	Brown
Females	White	1	6
	Brown	5	2

Determine if the null hypothesis of no association may be rejected. Note that some expected cell sizes are < 5.

7.16 We would like to test the efficacy of a drug papaveretum in treating acute abdominal pain. Based on the results, is administration of the drug associated with pain relief?

	Papaveretum	Saline
Pain Better	47	7
No change/worse	3	42

Data from Attard et al. 1992. *BMJ* 305: 554–556

7.17 An epidemiologist investigating an outbreak of food poisoning (gastroenteritis) at a church picnic used two steps to determine the cause. She first surveyed everyone attending the picnic to determine which food(s) is most likely to have caused the illness. (She later confirmed the cause by sampling food and ill people and culturing the bacteria.) To survey, she asked each person whether or not they became ill (vomiting and/or diarrhea within 24 hours of the picnic) and specifically which foods they ate and did not eat. Compiling information from 66 respondents allowed her to generate the following data (modified from Lilienfeld and Lilienfeld, 1980). Use a chi-square test of independence to determine if each of the foods is linked to illness. Note that you will not be able to test the first food (chicken) because only one person did not eat the chicken.

	Ill	Not Ill	Total	Attack rate (%)
Chicken:				
Eaten	49	16	65	75
Not Eaten	0	1	1	0
			difference	75
Chicken Dressing:				
Eaten	37	5	42	88
Not Eaten	11	11	22	50
			difference	38
Potato Salad:				
Eaten	37	5	42	88
Not Eaten	11	9	20	55
			difference	33
Cole Slaw:				
Eaten	27	7	34	79
Not Eaten	17	9	26	65
			difference	14

(handwritten: 44, 44 16, 64)

7.18 Locate a set of frequency data either from the popular press or from government websites, such as the Centers for Disease Control (http://www.cdc.gov/mmwr/). Generate a contingency table. Next, state the hypothesis you are testing. Then, conduct a chi-square test, reporting both the statistic and p-value, make a decision about your hypothesis, and report a verbal conclusion.

CHOOSE THE CORRECT TEST

7.19 In order to study the dependence of hypertension on smoking habits, the following data were collected from 160 individuals. Test the hypothesis that the presence or absence of hypertension is independent of smoking habits.

	Smoking habit		
	non-	moderate	heavy
Hypertension	21	36	30
No hypertension	28	26	19

7.20 In snapdragons, red flower color is incompletely dominant. Homozygous dominant individuals are red, heterozygous individuals are pink, and homozygous recessive individuals are white. In a cross of 2 heterozygous individuals, a ratio of 1 red to 2 pink to 1 white is expected in the offspring. The results of such a cross were 10 red, 21 pink, and 9 white. Do the observed results differ significantly from a 1:2:1 ratio?

7.21 Forty-seven groups of common suckers, each consisting of 3 individuals, were surveyed during their spawning season. The sex ratio in the population may be assumed to be 1:1. The number of males and females in each group of 3 were as follows.

Males	Females	Frequency
3	0	7
2	1	35
1	2	3
0	3	2

Is it likely that the number of males and females in groups of 3 individuals are due to chance alone?

7.22 Based on our understanding of X and Y chromosomes, it is expected that female births equal male births in most mammalian species. In a sample of 60 individuals from a population, 28 were males, and 32 were females. May we conclude that the sex ratio in this population is something other than 1:1?

7.23 As part of a conservation biology study, a researcher used radio transmitters to locate winter roosts of the red bat (*Lasiurus borealis*). The following data show the number of trees containing roosts compared with randomly chosen trees which lacked roosts. May we conclude that there is an association between bat roosts and tree species?

Tree species	With roosts	Without roost
Cedar (*Juniperus*)	11	6
Oak (*Quercus*)	10	15
Other	0	21

7.24 We suspect that a certain strain of laboratory rats has a genetic tendency to make left turns in a "T" maze. Of 12 rats that were tested in such a maze, 8 chose to go into the left arm and

4 chose the right arm. Does this result support our suspicion about a left-turning tendency?

7.25 One of the largest health experiments ever conducted investigated the effectiveness of the Salk vaccine in preventing paralysis and death from poliomyelitis. The poliovirus was a devastating disease because it mostly affected young children. The study generated the results below. Test the null hypothesis that the treatment and rate of polio are independent.

Treatment	Number with Polio	Number without Polio
Salk vaccine	57	200688
Placebo control	142	201087

Data modified from Meier 1989

Notes

[1] **Proof**

Let A represent R_i and B represent C_j and $E(A \wedge B)$ represent E_{ij} (as shown in table 7.4)

If independent, $P(A \wedge B) = P(A)*P(B)$

$P(A) = A/N$

$P(B) = B/N$

For cell $A \wedge B$, $E(A \wedge B) = P(A)*P(B)*N = (A/N)*(B/N)*N = (A*B)/N$

Since $A = R_i$, $B = C_j$, and $E(A \wedge B) = E_{ij}$

$E(A \wedge B) = (A*B)/N$ can be replaced by $E_{ij} = (R_i*C_j)/N$ (equation 7.3)

chapter eight

The Normal Distribution

it feels great Motherfucking yes

The **normal distribution** is a continuous probability distribution that closely describes a wide variety of continuous measurement variables, such as heights, weights, and agricultural yields. This close fit is because many such traits are the result of many factors (various genes, nutrient levels, etc.) working together independently. Many discrete variables also approximately follow such a distribution when the range of values that the variable may assume is fairly large. We can also use the distribution to easily determine probabilities. The normal distribution underlies a number of important statistical procedures, including *t* tests, ANOVA, and regression analysis. The normal distribution thus has many applications in statistics. In this chapter, we will explore some properties and uses of the normal distribution.

8.1 The Normal Distribution and Its Properties

Continuous measurement variables can assume any value between certain limits. For example, recall the mosquito fish lengths in table 3.3. An individual fish 25 mm long could, if measured with a fine enough ruler, actually be 25.01253274 mm. The histograms we generated from 172 measurements (fig. 3.2) were informative, but somewhat crude. Imagine if we instead had collected and measured a very large number of fish (say a million). If we then divided these measurements into many bins, each very thin, the histogram would become a smooth curve, somewhat like figure 8.1. The probability distribution that describes curves of this general type is the normal distribution.

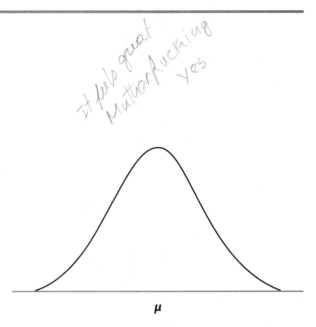

Figure 8.1 A normal distribution, centered at the population mean, μ. Since the normal curve extends to ± ∞, the curve approaches the *x* axis as an asymptote.

The normal probability distribution has certain important characteristics, which are discussed below and illustrated in figures 8.2 and 8.3.

Properties of the Normal Distribution

1. The distribution is completely defined by the mean (μ) and the standard deviation (σ). The location of the curve along the *x* axis is defined by the mean, and the spread of the curve is defined by the standard deviation (fig. 8.2). Since these parameters can potentially assume any value, there are an infinite number of normal distributions.

2. The height on the *y* axis of any point along the curve (*x*) represents the probability density function, *f(x)*, of that value of the variable, and the areas under the curve are determined with integral calculus.[1] Fortunately for us, these areas are readily available in tables.

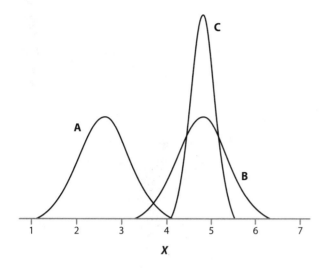

Figure 8.2 Three normal distributions. Curves A and B have the same variance, but different means; curves B and C have the same means, but the variance of B is larger than that of C. [After Sokal and Rohlf 1995]

3. The curve is perfectly symmetrical about the mean. Thus, if we know the probability for a section of one side, we know what the probability is for a similar section on the other side of the distribution.

4. One standard deviation above the mean includes 34.13% of all of the individuals in the population, and one standard deviation below the mean includes 34.13% of all of the individuals in the population. Thus, 68.26% of all of the individuals in the population fall within plus or minus one standard deviation of the mean (fig. 8.3). Another way of thinking about this is that there is a probability of 0.6826 that any individual (*x*) taken at random from this population would fall within one standard deviation of the mean. In a similar manner, the mean ±2 standard deviations contains 95.46% of all of the individuals in the popu-

lation and the mean ±3 standard deviations contains 99.73% of all the individuals within the population (table 8.1). We will have occasion to examine this property in considerably more detail in this and subsequent chapters.

Table 8.1 Proportions of the Area under the Normal Curve as a Function of the Standard Deviation

Standard Deviations	Proportion of Area
± 1	0.6826
±1.960	0.9500
± 2	0.9546
± 2.576	0.9900
± 3	0.9973

The total area under the normal curve is, by definition, exactly 1. Since the normal curve is defined by the mean and the standard deviation, there is a fixed relationship between the standard deviation and the proportion of the area under the curve that occurs between the mean and any standard deviation "unit." The relationships shown in table 8.1 hold true for any normal distribution. We will later use an expansion of this table to find areas more exactly.

8.2 The Standard Normal Distribution and Use of Z Scores

Fortunately, we do not have to deal with complex computations with the normal probability density function every time we wish to do something with the normal distribution. Among the infinite variety of normal distributions that are possible, there is one to which all others may be easily converted: the **standard normal distribution**. It has a mean of zero and a standard deviation of one, and its "tails" extend from minus infinity (wherever that might be) through zero (the mean) to positive infinity. Table A.1 in the appendix gives the proportion of the standard normal curve that lies between zero (the mean) and almost any value of a variable that one might reasonably expect to encounter. How this table is used is discussed in the examples that follow.

We often need to determine where, on a normal curve, a particular value falls, or more often, the probability that some value or range of values lies within a certain portion of the curve. To do this, individual variants (measurements) need to be expressed as if they were variants of the standard normal distribution. In other words, we need to

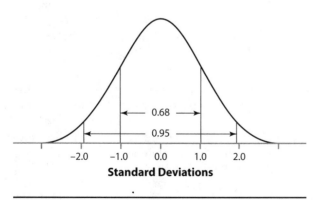

Figure 8.3 Areas of a normal distribution.

convert our normal distribution, which may have any imaginable mean and standard deviation, to the standard normal distribution, which has a mean of zero and a standard deviation of one. In practice it is unnecessary to convert all of our observations; only those of interest need be converted. These converted observations are called **z scores**, standard scores, or standard normal deviants. Equation 8.1 transforms any variable into its corresponding z score.

$$z = \frac{x - \mu}{\sigma} \qquad (8.1)$$

What might not be immediately apparent here is that a z score is how far from the mean, in terms of standard deviations, an observation lies. An observation with a z score of +1 is one standard deviation greater than the mean, and an observation with a z score of –1 is one standard deviation less than the mean (fig. 8.3). An observation with a z score of 0 has the same value as the mean. Thus, any variant is expressed not in its actual value but in how far, in terms of standard deviations, it lies from the mean.

Example 8.1
An Approximately Normally Distributed Variable

The height (in centimeters) of a large group ($n = 414$) of female general biology students was determined. We will consider this group to be the entire population in which we have an interest. The mean height of this group was 166.8 cm, and the standard deviation was 6.4 cm.

Height in humans is an approximately normally distributed variable. We will use the standard normal distribution to answer several questions about this population.

Question 1. What is the z score of an individual who is 170 cm tall?

$$z = \frac{x - \mu}{\sigma} = \frac{170 - 166.8}{6.4} = 0.5$$

It is easy enough to comprehend that this person is 3.2 cm taller than the average student in this population, so why is it of interest to us that she is 0.5 standard deviations taller than the average? While we ponder that, consider another question.

Question 2. What proportion of the population is as short as or shorter than this 170 cm tall person? A somewhat different way of asking this same question is this: what is the probability that an individual sampled at random from this population will be 170 cm or shorter? Keep in mind in this and

subsequent discussions that the area of a probability distribution corresponding to a designated proportion of the distribution may also be thought of as a probability. To solve this problem, we need to become acquainted with table A.1. When considering the use of table A.1 in this example, it will help to refer to figure 8.4.

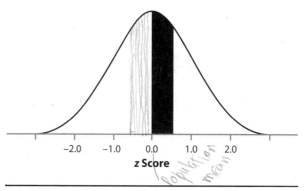

Figure 8.4 The standard normal distribution (for example 8.1, question 2).

Zero on the x axis of this graph represents a z score of 0, which is the population mean. The shaded area of the curve (the area between the mean and our z score) represents the proportion of the standard normal distribution that falls between the mean and the indicated value of z, and these are the entries in table A.1. The left column of the table gives z scores to one significant digit. The top row extends this to 2 significant digits. For example, if we wish to know the proportion of the curve that lies between the mean and a z score of 1.23, we look down the left column to $z = 1.2$ and go across to the column headed .03 (thus, 1.23). The value we find there is 0.3907. Look this up yourself in appendix A right now.

Let us return to the original question, but rephrase it: what proportion of the standard normal distribution lies less than a z score of 0.5? We first need to determine what proportion of the standard normal distribution lies between the mean and a z score of 0.5 (fig. 8.4). According to table A.1, the area is 0.1915 (verify this for yourself). Therefore, 0.1915 (or 19%) of the individuals in this population lie between 170 cm (our person with a z score of 0.5) and 166.8 cm, which is the mean. Note that the standard normal distribution is perfectly symmetrical and that 0.5000 of the area lies above the mean and 0.5000 of the area lies below the mean. Our original question was: what proportion of the population is as short as or shorter than 170 cm? We know that 0.5000 (50%) lies below the mean (by definition). We also know that 0.1915 (19%) lies between the mean and a z score of 0.5 (corresponding to 170 cm). Therefore, 0.5000 + 0.1915, or 0.6915 (roughly 69%), lies below a z score of 0.5. We conclude that the proportion of the individuals in this population who are 170 cm or shorter is 0.6915, or in other words, approximately 69% of the individuals in this population are as short as or shorter than this individual who is 170 cm tall. This may also be interpreted to mean that the probability that any individual sampled from this population will be 170 cm or shorter is 0.6915.

You may be curious about why the 0.5000 of the standard normal curve that lies between the mean and negative infinity is not already added to the entries in table A.1. Another question might help to clarify this.

Question 3. What proportion of this population would be expected to be 163.6 cm or shorter? The z score for this observation is:

$$z = \frac{163.6 - 166.8}{6.4} = -0.5$$

Graphically, the problem looks like figure 8.5. We are trying to determine how much (what proportion) of the standard normal distribution lies to the left of (below) our z score of –0.5. Now since the standard normal distribution is symmetrical, the same proportion lies between the mean and a z score of –0.5 that lies between the mean and a z score of +0.5, which we already know is 0.1915. (If you do not remember how we already know this, refer back to the last question.)

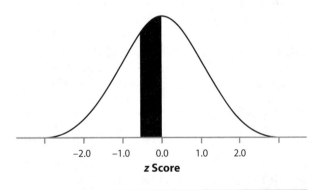

Figure 8.5 The standard normal distribution (for example 8.1, question 3).

Thus, the proportion that lies below a z score of –0.5 is:

$$0.5000 - 0.1915 = 0.3085$$

We conclude that approximately 31% of this population will be as short as or shorter than 163.6 cm, or that the probability that any individual sampled at random from the population will be 163.6 cm or shorter is 0.3085.

Question 4. What proportion of the population lies between 160 cm and 170 cm? This problem is shown graphically in figure 8.6. To solve it we first determine what proportion of the curve lies between the mean and 160 cm, which is:

$$z = \frac{160 - 166.8}{6.4} = -1.063$$

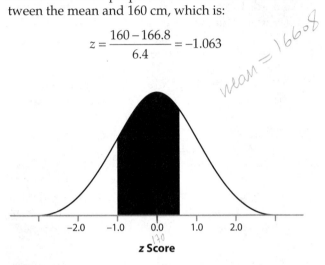

Figure 8.6 The standard normal distribution (for example 8.1, question 4).

According to table A.1, 0.3554 of the standard normal curve lies between the mean and a z score of –1.063 (–1.06, actually, since the table only goes to 2 places for z scores). Next we determine how much of the curve lies between the mean and 170 cm, for which the z score is:

$$z = \frac{170 - 166.8}{6.4} = 0.50$$

Again, according to table A.1, 0.1915 of the standard normal distribution lies between the mean and a z score of 0.50. Thus, the proportion lying between 160 cm and 170 cm, or between a z score of –1.06 and 0.50, is 0.3554 + 0.1915, or 0.5469. Thus, approximately 55% of this population is expected to consist of individuals whose height is between 160 cm and 170 cm. The probability that any individual sampled at random from this population will be between 160 cm and 170 cm is 0.5469.

Question 5. What range of heights includes 0.95 (95%) of this population? This problem is somewhat different from the previous ones. Here we know the proportion of the standard normal distribution in which we have an interest (that is, we know what the z score is, and we must find the value of x that corresponds to this z score). Table A.1 shows that 0.475 of the standard normal distribution lies between the mean and a z score of 1.96, and therefore, 0.475 lies between the mean and a z score of −1.96. Thus, 0.95 is between the mean and a z score of ±1.96. (Refer back to figure 8.3 to help visualize this situation.) Now since

$$z = \frac{x - \mu}{\sigma}$$

it follows that

$$x = z\sigma + \mu$$

Accordingly, the height (x) that includes 0.475 of the standard normal distribution below the mean is

$$x = (-1.96 \times 6.4) + 166.8 = 154.23$$

and the height (x) that includes 0.475 of the standard normal distribution above the mean is

$$x = (1.96 \times 6.4) + 166.8 = 179.34$$

We conclude that 95% of the individuals in this population are between the heights of 154.23 cm and 179.34 cm. Or, thinking of probabilities, there is a probability of 0.95 that any individual sampled at random from this population will be between 154.23 cm and 179.34 cm, and a probability of 0.05 that any individual sampled at random from this population will be taller than 179.34 cm or shorter than 154.23 cm.

Since we rarely know the population mean and standard deviation of a population, of what use is the information that we have been considering? From an applied standpoint, not much. However, an understanding of these concepts gives us the background to become acquainted with another group of probability distributions, called sampling distributions, and further insight into using probabilities for statistical inference.

8.3 Sampling Distributions

Consider an example. A random sample of 10 bluegill sunfish standard lengths was selected. The mean of this sample was 159.40. Based on this sample, we would like to conclude something about the true population mean (μ). The value of this parameter is usually unknown to us. Using the sample mean, we get a view through a particular random sample that we happen to collect.

When a random sample is taken from a population, the sample that was drawn was only one of a very large number of samples of the same size that could have been taken (fig. 8.7). If we were to take repeated random samples of the same size from a

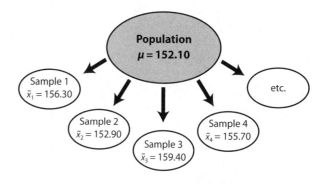

Figure 8.7 Graphic representation of repeated samples from a population, such as the bluegill sunfish example.

normally distributed population and compute the mean of each sample, we would likely find these means to be different from each other. Furthermore, these means would have values tending to cluster about some central value. In other words, these means would have a mean and standard deviation of their own and would exhibit all of the characteristics of a normally distributed random variable in exactly the same way that observations from a normally distributed population do. It is completely appropriate to think of a sample mean as an individual observation taken from a large population of possible sample means. The probability distribution that describes the behavior of a repeated sample statistic (mean, standard deviation, and so on) is called a **sampling distribution**.

8.3.1 The Central Limit Theorem

In the case of sample means taken from a normally distributed population with a known population standard deviation, the sampling distribution of these means also follows a normal distribution centered at the true population mean (μ). Interestingly, even for non-normal populations, if sample sizes are large enough their means will tend to also follow a normal distribution. This property is so useful for statistical inference that it goes by the formal name of a theorem.

> **The Central Limit Theorem**
>
> The means of samples from a population with a non-normal distribution will approximate a normal distribution as long as the sample size is large enough.

Figure 8.8 illustrates the central limit theorem. Here, the underlying distribution of city sizes is skewed, with most being small. Let's say we collect

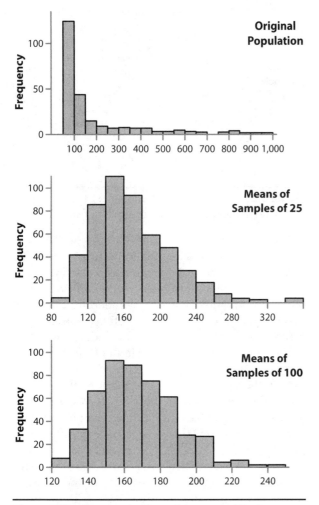

Figure 8.8 Demonstration of the central limit theorem. The original population is the population size (in thousands) of cities between 50,000 and 1 million people. (From Snedecor and Cochran 1989)

repeated samples of size n, compute the sample mean for each, and then plot the frequency distribution of the means. The distribution of means are less skewed and become approximately normal as sample size gets large.

The useful thing about the central limit theorem is that we do not have to repeatedly take samples from a population to know how the means of such samples would behave. These means would have a normal distribution. Thus, we may consider a single sample mean from a normally distributed population (or from a non-normally distributed population, if our sample size is large) as one of many possible means whose distribution would be normal.

8.3.2 Standard Error of the Mean

Another important aspect of this sampling distribution is that, as n grows large, the distribution of

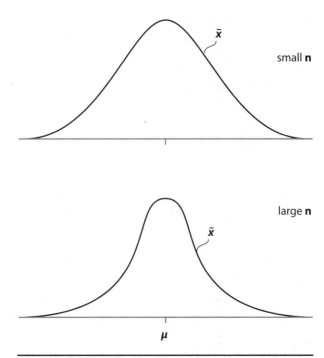

Figure 8.9 Sampling distribution of means from the same population. Sample means from large samples (bottom) have a greater chance of being close to the true mean than sample means from small samples (top).

means becomes narrower in shape, indicating that most sample means are close to the true mean μ. In other words, our precision for estimating the mean (or any other parameter) improves as n grows larger. This property is illustrated in figure 8.9.

We may assign a z score to a sample mean if we know (or are willing to assume) the population mean (μ) and standard deviation (σ) of the population from which our sample was taken. Recall from equation 8.1 that a variant of the standard normal distribution for a single observation is calculated as $z = (x - \mu)/\sigma$:

Example 8.2

Assume that the bluegill sunfish whose lengths are given in Digital Appendix table 1 are the entire population of interest. (Imagine that we had completely drained a pond.) The population mean length (μ) for this population is 152.10 mm, and the population standard deviation (σ) is 19.64 mm. Suppose that we have taken a random sample of 10 individuals from this population (fig. 8.7) and obtained a sample mean $\bar{x}$ of 159.40 mm.

We may now ask this question: What is the probability of obtaining a sample mean of 159.40 mm or larger from a population whose population mean is 152.10 mm and whose population standard deviation is 19.64 mm? We may answer this question in much the same way that we answered questions about the height of women students in example 8.1, by assigning a z score to this sample mean. This problem is illustrated graphically in figure 8.10.

Figure 8.10 The standard normal distribution. Shaded area is the proportion above a z score of 1.18.

The z score for a sample mean is given by

$$z = \frac{\bar{x} - \mu}{\sigma_{\bar{x}}} \quad \text{(8.2)}$$

The new term in this equation, $\sigma_{\bar{x}}$, is the standard deviation of our theoretical population of sample means; it is called the **standard error of the mean**, and it is calculated by

$$\sigma_{\bar{x}} = \frac{\sigma}{\sqrt{n}} \quad \text{(8.3)}$$

where σ is the population standard deviation and n is the sample size in question. Note that, in this situation, the parameter σ is known.

Notice how the standard error of the mean (equation 8.3) changes with sample size. As n grows large, $\sigma_{\bar{x}}$ becomes small. This property is illustrated in figure 8.9. In other words, sample means from large samples are more likely to be close to the true population mean than those of small samples. We have greater precision of our estimate.

For the example above,

$$\sigma_{\bar{x}} = \frac{19.64}{\sqrt{10}} = 6.21$$

The z score for our sample is therefore

$$z = \frac{159.40 - 152.10}{6.21} = 1.18$$

We could, of course, calculate this more simply by combining the two equations above to give

$$z = \frac{\bar{x} - \mu}{\sigma / \sqrt{n}} \quad \text{(8.4)}$$

Consulting table A.1 we find that 0.3810 of the standard normal distribution lies between the mean and a z score of 1.18. Subtracting this value from 0.5000 gives 0.1190 of the standard normal distribution that lies above a z score of 1.18 (the shaded portion of fig. 8.10). Thus, the probability of drawing a sample of 10 individuals with a mean as high as or higher than 159.40 mm is 0.1190. Another way of saying this is that approximately 12% (0.1190 × 100) of all of the samples of 10 that might possibly be drawn from this population would have a mean of 159.40 or higher.

Compare this answer to the probability of obtaining any single individual with a length greater than or equal to 159.40 mm. Here, z = (159.40 − 152.10)/19.64 = 0.37. After consulting table A.1, we find $P(x \geq 159.40) = 0.5000 - 0.1143 = 0.3857$. The probability here (39%) is considerably greater than that for mean being so large. In general, the probability of getting extreme outcomes is greater for an individual measurement than it is for the mean.

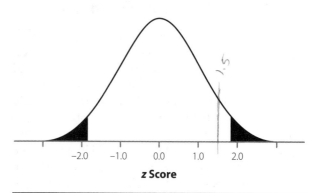

Figure 8.11 The standard normal distribution. Shaded areas represent 0.05 of the distribution.

There is another question we might ask, and one that is more to the point: Within what range would we expect 95% of the sample means of samples of 10 to lie? This is equivalent to the question asked in example 8.1 regarding the range of heights that would include 95% of the population of female freshman biology students. Refer to figure 8.11. We are looking for those values of sample means that include the unshaded portion of this curve, or the values

$$\mu + z_{(0.475)}\sigma_{\bar{x}} \quad \text{(8.5)}$$

and

$$\mu - z_{(0.475)}\sigma_{\bar{x}} \qquad (8.6)$$

Recall that 0.95 of the standard normal distribution lies between z scores of -1.96 and 1.96 (see table A.1). For the example these values are:

$$152.10 + (1.96 \times 6.21) = 164.27$$

and

$$152.10 - (1.96 \times 6.21) = 139.93$$

Thus, 95% of the samples of 10 that we might draw from this population would have a sample mean between 139.93 and 164.27. In other words, there is a probability of 0.95 that any sample of 10 we might draw from this population would have a sample mean between 139.93 and 164.27. Conversely, there is a probability of 0.05 that any sample mean would lie outside of this range—either smaller than 139.93 or larger than 164.27 (the shaded portion of fig. 8.11).

In chapter 9, we will investigate this concept under the more realistic scenario where we don't know the actual values of the population parameters when we compute the confidence interval for μ.

8.4 Testing for Normality

Many of the statistical tests that we will consider in the following chapters are based on the assumption that the variable in question is at least approximately normally distributed. This is no accident. The development of statistics during the early 1900s occurred before computers became widely available and many of these procedures are in principle quite simple mathematically. However, one should never assume that a variable is normally distributed unless there is some reason to believe that this is the case. Example 8.1 asserted that height in humans is an approximately normally distributed variable. How do we know this? One way is to examine the frequency distribution of the variable from a large sample.

Consider the male heights given in exercise 3.1. The histogram for these data, using a class interval of 2 cm, is shown in figure 8.12. Note that this histogram is more or less bell-shaped and its overall appearance is much like that of a normal distribution. A better way to check for normality is to use a probability plot.

A **probability plot** is a graph of the cumulative frequency (section 3.4) plotted on the y axis against the original measurement variable (on the x axis),

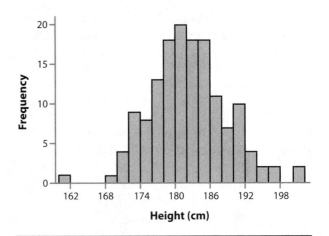

Figure 8.12 Frequency distribution of human male heights (data from exercise 3.1).

where the y axis uses a normal probability scale. The y axis is scaled in such a way that, when data are drawn from a normal distribution, the points will fall close to a straight line. Such a plot is particularly useful because the human eye can readily detect departures of points from a straight line, easier than from a curve. An example of such a plot is shown in figure 8.13. Notice that the plotted points are quite close to the straight line, suggesting that this sample of human heights was taken from a population having a normal distribution. Many statistical packages contain programs for doing normality tests and s. For example, the "normality test" routine in *Minitab* generates the plot and also does a type of goodness of fit test to formally check for normality. In figure 8.13, the Anderson Darling test ("AD") shows a p-value of 0.201, consistent with a close fit of the data to a normal distribution.

8.5 Normal Approximation of the Binomial Distribution

Numerous biological characteristics are binary, categorical variables with only two possible classes. Subjects are dead or alive, male or female, mutant or wild type, etc. When we studied the binomial distribution (section 5.2), we arbitrarily called one of these classes a success (with probability p) and the other a failure (with probability $q = 1 - p$). If we collect a sample of size k, we can use the binomial equation to determine the probability that x takes on any specific value. (In selecting 5 individuals, what is the probability that exactly 2 are female?) Recall also that we are often interested in knowing a range of probabilities (cumulative probability),

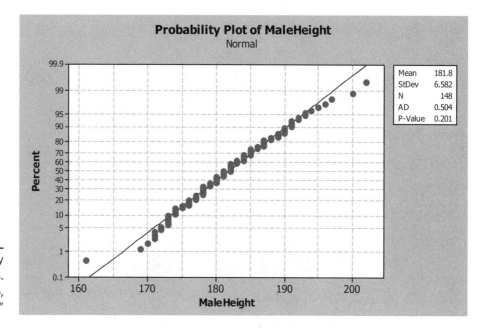

Figure 8.13 Probability plot for human male heights. Generated from Minitab 16, using "normality test."

something we will apply when making inferences about proportions (section 8.6 below).

For now, let's think about how best to determine cumulative probabilities. We can calculate the individual binomial probabilities and add them up (section 5.4). You may recall that doing many of these calculations can be quite tedious with a hand calculator, but more easily solved with a computer. Knowing the central limit theorem gives us another approach that will frequently be handy. When k is fairly large and p is not too near 0 or 1, the binomial distribution becomes approximately a normal distribution. How large is "fairly large"? As a rule of thumb, if the products $k*p$ and $k*q$ are both at least 5, the normal approximation is considered to be fairly accurate.

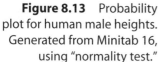

Rule
To use the normal approximation to the binomial distribution, each of the products $k*p$ and $k*q$ must be ≥ 5.

Example 8.3

Suppose that 25 individuals are sampled from a population in which the ratio of females to males is known to be 1:1, and we wish to know the probability of obtaining a sample of eight or *fewer* males in this sample of 25 individuals. Let p = the probability that any individual sampled will be a male (0.5) and thus q = the probability that any individual sampled will be a female (0.5), and let x = the number of males in which we have an interest (8 or fewer, in this case). The entire probability distribution is illustrated in figure 8.14 on the following page.

One approach is to use the Binomial formula to calculate each of the nine probabilities (x = 0, 1, 2, ..., 8) and then sum each of these answers to get our desired result, the probability of $x \leq 8$. Using *Minitab* to solve the cumulative probability gives the answer as: $P(x \leq 8)$ = .0322. However, we can also solve the problem using the normal approximation to the binomial distribution. Notice the word "approximation." The answer from using the normal distribution won't be exactly right, but often the answer will be close enough.

The values of $k*p$ and $k*q$ are both equal to 12.5 (and thus greater than 5, satisfying the rule above). We can therefore feel comfortable in approximating this binomial distribution with the normal distribution.

The mean of a binomial variable is given by

$$\mu = kp \tag{8.7}$$

which, for the example, is

$$\mu = 25 \times 0.5 = 12.5$$

or 12.5 males out of 25 individuals.

The standard deviation is given by

$$\sigma = \sqrt{kpq} \tag{8.8}$$

which, for the example, is

$$\sigma = \sqrt{25 \times 0.5 \times 0.5} = 2.5$$

We may regard the possible outcomes of the various combinations (no males and 25 females, 1 male and 24 females, and so on) as a normally distributed variable with a mean of 12.5 and a standard deviation of 2.5. Reference to figure 8.14 might

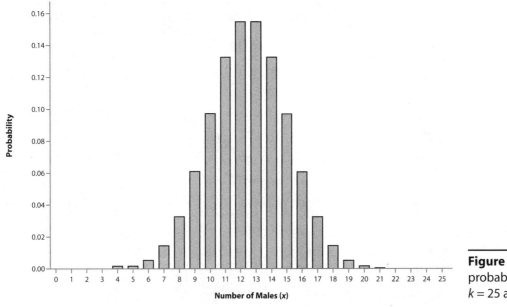

Figure 8.14 Binomial probability distribution for $k = 25$ and $p = 0.5$.

help in visualizing this situation. The bars in the figure represent the binomial probability distribution for this example. Note that the general shape of this distribution is very similar to that of a normal distribution except that it is a bar graph. Equation 8.1 is now used to compute the z score for a value of 8, given that the mean is 12.5 and the standard deviation is 2.5.

$$z = \frac{x - \mu}{\sigma}$$

The value 8 has a z score of

$$z = \frac{8 - 12.5}{2.5} = -1.80$$

In table A.1 we find that 0.4641 of the normal curve lies between the mean and $z = 1.80$ (and therefore, that 0.4641 lies between the mean and −1.80). Since 0.500 lies above the mean, the region lying beyond a z score of −1.80 is

$$0.5000 - 0.4641 = 0.0359$$

In other words, the probability of obtaining a sample of 25 individuals in which there are eight or fewer males is 0.0359. This is in pretty close to the answer computed by *Minitab* for exact binomial cumulative probabilities (.0322). When n is larger, the answers would be closer still.

This technique provides only an approximation of the binomial probability distribution, since the binomial distribution deals with discrete variables and the normal distribution deals with continuous variables. The discrepancy between the two is not large when k is large (30 or so), but it may become

appreciable when k is smaller. For these smaller samples, we can use a "continuity correction"[2] to get a closer approximation. However, for these samples it's usually more convenient just to compute the cumulative binomial probability directly using available computer software.

8.6 Using the Normal Approximation for Inferences about Proportions

Statistical inference refers to making a conclusion about a population based on a sample taken from that population. Recall that there are two types of inference, estimation and hypothesis testing. We can make inferences about a population proportion. We can calculate a confidence interval, with some probability (often 95%) and we can test hypotheses.

8.6.1 Confidence Interval for a Proportion

Earlier in section 6.1, we considered a situation of collecting a sample of 100 frogs and finding that 15% of them were infected with the chytrid fungus (prevalence = 0.15). If we want to infer the infection prevalence for the entire population of frogs, p, the best estimate is $\hat{P} = 0.15$. (This estimate can be called "P_{hat}.") Because this value was taken from a sample, subject to random chance, p is unlikely to be exactly 0.15. Instead, the true value of p is likely (with 95% certainty) to be within the range 0.089–0.234. How did we determine that range? To calculate the confidence interval, first determine the standard error. One can calculate the standard error of any proportion, using its estimate as:

$$SE_{\hat{p}} = \sqrt{\frac{\hat{P}\left(1-\hat{P}\right)}{n-1}} \qquad (8.9)^3$$

The confidence interval is then determined by multiplying this standard error term by a critical z value (from table A.1) and then either subtracting or adding this product to $\hat{P}$. For a 95% confidence interval, the critical z value is 1.96, since this value cuts off 5% of the distribution in the two tails (see question 5 in section 8.2 above). If a different confidence interval is desired (say 99%), simply choose a different critical z value. The 95% confidence for p is:

$$\hat{P} - \left(1.96 * SE_{\hat{p}}\right) < p < \hat{P} + \left(1.96 * SE_{\hat{p}}\right) \qquad (8.10)$$

Another approach is to simply use a table. Refer to table A.11 for confidence limits for percentages. Sample sizes are indicated on the top row and estimated proportions ($\hat{P}$) are shown in the left-hand column. For instance, in our earlier frog example, the estimated proportion infected from a sample of 100 frogs was 0.15. Checking the table, the lower and upper limits for the 95% confidence interval is 0.089 and 0.234, respectively. Inspect the table. Notice that, had a larger sample of frogs been collected, the confidence interval would have been narrower. This illustrates the fact that precision improves as sample sizes grow large.

8.6.2 Binomial Tests

In chapter 7, we used chi-square to test hypotheses about frequency data. We used a goodness of fit to test how well a single sample conformed to a particular probability model and we used a test of independence to compare the frequencies among classes in different groups. Chi-square is the most common procedure for testing these hypotheses. There are times when a binomial test is useful as an alternative statistical procedure, such as when the hypothesis is "one-sided."

For instance, consider the 1954 experimental trial of the Salk polio vaccine (table 8.2). The researchers wanted to compare the prevalence of polio in two groups of patients, one treated with the Salk vaccine and the other treated with a placebo control. The experiment was conducted with the following results:

Table 8.2 Results from the Salk Polio Vaccine Trial (modified from Meier 1989)

Treatment	Number with Polio	Number without Polio
Salk vaccine	57	200688
Placebo control	142	201087

The usual procedure is to run a chi-square test of independence. Refer back to exercise 7.25. You should have found that chi-square statistic is 36.12, with $p < 0.001$. Clearly, the null hypothesis is rejected, indicating that there is a significant association between treatment and disease state. We can view the hypothesis a different way. This chi-square test of independence for these two groups is equivalent to testing the null hypothesis:

$H_0: p_c = p_v$ versus

$H_A: p_c \neq p_v$

where p_c and p_v refer to prevalence of polio in the control and vaccine groups, respectively. The true prevalence in each group (p) is estimated by the proportion with polio ($\hat{P}$), determined by dividing the number with polio by the total number receiving that particular treatment.

In this experiment, our hypothesis is a bit more narrow than that stated above. The question of interest (our working hypothesis) concerns the vaccine *lowering* the prevalence of polio. In other words, we expect those receiving the placebo control to have a higher prevalence than those receiving the vaccine. Thus, the statistical hypotheses become:

$H_0: p_c = p_v$ versus

$H_A: p_c > p_v$ control should see more polio

This is an example of a **one-sided hypothesis**, since we are only interested in rejecting the null hypothesis in one direction. Note that the null hypothesis could also be stated as $H_0: p_c \leq p_v$; either expression is correct. If the prevalence of polio turns out to be higher in the vaccinated group, we would still fail to reject H_0.

To test this hypothesis, we can conduct a binomial test using a Z statistic. We will reject the null hypothesis if the Z statistic exceeds the critical z value from table A.1. For a test at $\alpha = 0.05$, the critical value is 1.96. Check for yourself that this number correctly cuts off 5% of the distribution in the right hand tail. For two samples (1 and 2), the Z statistic is calculated as:

$$z = \frac{\hat{P}_1 - \hat{P}_2}{\sqrt{\dfrac{\hat{P}_1\left(1-\hat{P}_1\right)}{n_1} + \dfrac{\hat{P}_2\left(1-\hat{P}_2\right)}{n_2}}} \qquad \text{(8.11)}$$

The term in the denominator is the standard error of the difference between two proportions. For this particular example, $\hat{P}_v = 0.000284$ and $\hat{P}_c = 0.000706$ and the corresponding sample sizes are 200,745 and 201,229. Plugging the numbers in the formula, putting the control group in place of group 1, gives a value of $Z = 6.0133$. The corresponding p-value is very small ($p < 0.00003$). Refer to the table to confirm the answer.

When doing a one-sided test, it is very important to use the correct sequence in the numerator. The sequence in the formula should always follow the sequence in the null hypothesis. See our null hypothesis above to confirm that this example was done correctly.

Binomial tests are not frequently used. More common is to simply run chi-square tests. But as we saw they are sometimes handy and also demonstrate the utility of the normal distribution. For more extensive treatment of binomial tests, the reader should consult Whitlock and Schluter (2009).

Key Terms

central limit theorem
normal distribution
normality test
probability plot
sampling distribution
standard error of the mean
standard normal distribution
z scores

Exercises

8.1 Find the following probabilities using the normal distribution.
   a. P (Z > 1.43)
   b. P (Z > –1.43)
   c. P (0.19 < Z < 2.17)
   d. P (–2.21 < Z < 1.14)

8.2 Find the following probabilities using the normal distribution.
   a. P (Z < 0.23)
   b. P (Z < –1.67)
   c. P (–0.50 < Z < 1.00)
   d. P (–2.5 < Z < 2.5)

8.3 Answer the following questions based on a population of mosquito fish with a known mean length of 34.29 mm and a standard deviation of 5.49 mm.
   a. What is the probability that any individual sampled at random from this population would have a length of 40 mm or larger?
   b. What is the probability that a random sample of 10 individuals would have a mean length of 40 mm or larger?
   c. What is the probability that any individual sampled at random will have a length between 32 mm and 42 mm?

8.4 Answer the following questions based on a population of bluegills with a known mean length of 152.09 mm and a standard deviation being 19.62 mm.
   a. What is the probability that any individual sampled at random from this population would have a length of 140 mm or shorter?
   b. What is the probability that a random sample of 25 fish have a mean length of 140 mm or shorter?
   c. What is the probability that any individual sampled at random will have a length between 148 mm and 160 mm?

8.5 A study was conducted to see if there were racial differences in the birth weight of infants (gestation between 37 and 42 weeks) in a Northern California population. The mean weight of white infants was 3576 g with a standard deviation of 462 g.
   a. What is the probability that a white infant would be born with a weight of 3600 grams or larger?
   b. What weights include 95% (0.95) of the population of white infants.

(Data from Madan et al. (2002) *Journal of Perinatology* 22:230–235)

8.6 The weight of Chinese infants are on average smaller than that of white infants. In the same Northern California population, the mean weight of Chinese infants was 3397 g with a standard deviation of 422 g.
   a. What is the probability that a Chinese infant would be born with a weight of 3600 grams or larger?
   b. What weights include 95% (0.95) of the population of Chinese infants?

(Data from Madan et al. (2002) *Journal of Perinatology* 22: 230–235)

8.7 Using the data from digital appendix 1, use statistical software (e.g., *Minitab*) to construct a histogram and a probability plot of bluegill fish lengths. Use this information and any other procedures that your specific computer program has to decide if this variable is approximately normally distributed.

8.8 Using the data from digital appendix 3, use statistical software (e.g., *Minitab*) to construct a histogram and a probability plot of the pulse rate of males. Use this information and any other procedures that your specific computer program has to decide if this variable is approximately normally distributed.

8.9 According to the National Center for Health Statistics, as of 2012, it is estimated that 24% of adults in Missouri smoke. You wish to know the probability of selecting from this population a random sample of 30 people containing 8 or fewer smokers.

 a. Use a statistical program to determine the exact probability of getting 8 or fewer smokers. (Hint: refer to section 5.4.)

 b. Now use the normal approximation of the binomial distribution to answer the same question. Compare your answers. Are they similar?

8.10 A population of red-bellied snakes is known to have a ratio of grey color morph to red color morph of 53:47. You wish to know the probability of selecting a random sample of 20 snakes containing 3 or fewer red morph individuals.

 a. Use a statistical program to determine the exact probability of getting 3 or fewer grey morph individuals.

 b. Now use the normal approximation of the binomial distribution to answer the same question. Compare your answers. Are they similar?

8.11 The Northern Leopard Frog, which is easily distinguished by spots covering its body, is common throughout the United States and Canada. The *burnsi* morph of the Leopard frog is caused by a mutation in a single gene and is characterized by a lack of spots on the frog's back. In a sample of 500 frogs in Minnesota, 10% of the frogs were of the *burnsi* morph. Use

equation 8.10 to calculate the 95% confidence interval for this proportion and compare this interval to that found in table A.11.

8.12 In a sample of 1000 adults, 30% complain of suffering from insomnia. Use equation 8.10 to calculate the 95% confidence interval for this proportion and compare this interval to that found in table A.11.

8.13 The following results show the relationship between smoking in mothers and the survival of newborns past their first year. Use the binomial test to determine if the proportion of newborns that died in the first year is lower in nonsmokers than smokers.

	Non-smoker	Smoker
Died in 1st year	28	7
Alive at 1 year	5606	583

8.14 The following results show the relationship between snoring and heart disease. Use the binomial test and the following data to determine if people who snore every night are more likely to have heart disease than those who do not snore every night.

	Non-Snorers	Snores Every Night
Heart Disease	24	30
No Heart Disease	1355	224

Notes

[1] The probability density function is expressed formally as:

$$f(x) = \frac{1}{\sigma\sqrt{2\pi}} e^{-1/2(x-\mu/\sigma)^2}$$

We cannot think of this y axis value as representing a probability (or frequency) since a point has no dimension, and therefore, a line drawn vertically from such a point has no width. Thus, there is no real value that corresponds to this value of x. However, if we consider two values of x that are *very* close together, we could use integral calculus to compute the area of the curve that lies between these two values. In this way we may assign a probability to a very narrow *range* (class interval) of the variable in question.

[2] For a detailed explanation of the continuity correction, see Snedecor and Cochran (1980).

[3] This particular equation for the standard error of $\hat{P}$ is sometimes called the Wald method. There are various modifications of this method, none of them perfect (Whitlock and Schluter 2008), and so we have chosen to use the most common method.

Inferences about a Single Population Mean
One-Sample and Paired Comparisons

So far, we have examined the theory behind hypothesis tests (chapter 6) and used that process to test hypotheses about frequency data (chapter 7). In chapter 8, we explored the properties of the normal distribution and sampling distributions for the mean. We are now prepared to examine hypotheses about the mean, a topic covered in the next several chapters. In this chapter we test hypotheses about single means and use confidence limits to estimate the likely boundaries for the mean of a single sample. The same method can be extended to matched-pairs experiments and thus entry into exploring treatment effects.

9.1 Questions about the Mean from a Single Population

Quite often, biologists have questions about the true mean (μ) of some statistical population, based on a random sample from that population. What is the average nitrogen concentration in the soil of a farmer's field? What is the average blood pressure of some group of middle-aged Americans? How does blood pressure change following a particular medication? In each case, we don't know the exact value of μ, nor do we know the exact value of the population standard deviation (σ). We need to use inferential statistics to make conclusions about the population, based on a random sample from that population. A confidence interval can be constructed to estimate the likely boundaries for the mean and an appropriate one-sample statistical test can be used to evaluate a hypothesis about the mean.

We will start with a somewhat simple example to explore how procedures work. Later in the chapter (section 9.6), we will use the same procedure on differences in order to test questions about treatment effects in a matched-pairs study. Consider the following two examples.

Example 9.1
Hypothesis about a Single Population Mean

Vitamin X is an essential nutrient, but too much of it in the diet is harmful. Accordingly, the Food and Drug Administration (FDA) sets standards for the content of X in vitamin pills: they must contain an average of 100 units of the vitamin per pill. The manufacturer of vitamin pills assigns a biochemist to monitor the vitamin X content of their product. She selects a random sample of 50 pills from a particular lot and finds that the vitamin X content in that sample has a mean of 100.5 units, with a standard deviation of 2.19 units. The biochemist must decide if it is reasonable that a sample with a mean of 100.5 could have been drawn from a population with a mean of 100.

Example 9.2
Hypothesis about the Effect of Treatment in a Matched-Pairs Study: Burn Healing

Many people receive serious burns every year and the medical community continues to search for new treatment options. A new medication (N) was tested to see if it improves the rate of skin healing over that of the most commonly-used medication (O). A test was conducted on 20 patients who had severe burns on both arms. For each patient, one arm was randomly chosen to receive medication N and the other received medication O. Fol-

(continued)

lowing the course of treatment, the time (in days) necessary for wound healing was measured on each arm. The average difference (O – N) in the wound healing rate was +2.7 days with a standard deviation of 2.36 days. The doctor must decide if the difference represented a significant improvement of the new medication over the old medication.

9.2 The *t* Distribution

Before we determine confidence intervals about the mean, we need to learn about another important probability distribution. When we make inferences about the population mean, we generally do not know the exact value of the population standard deviation and instead rely on the sample standard deviation as an estimate of this parameter. When the population standard deviation is unknown, the sampling distribution of means departs somewhat from normality, and this departure becomes more extreme as sample size becomes smaller. To account for this, we use the *t* **distribution**, or as it is sometimes known, Student's *t* distribution.[1]

Student's *t* distribution is very similar to the normal distribution in most respects, except that it is defined by the **degrees of freedom** (df) in addition to the mean and standard deviation. For a single sample of size *n*, the degrees of freedom are simply computed as: df = *n* – 1. Recall that the proportion of the standard normal distribution that occurs between the mean and any *z* score may be determined by reference to table A.1. We may also determine the value of *z* that includes a specified proportion of the standard normal distribution by using table A.1. For example, a *z* value of ± 1.96 includes 0.95 (or excludes 0.05) of the standard normal distribution.

When considering the distribution of sample means, it is useful to think of *t* as having the same general properties as *z*, except that the values of *t* that exclude a given proportion of the *t* distribution vary with sample size. We may determine the value of ± *t* that excludes certain specified proportions of the *t* distribution by reference to table A.2.

Suppose we wish to determine what value of *t* excludes 0.05 of the *t* distribution with 4 degrees of freedom. A graphic representation of the *t* distribution is shown with table A.2. The shaded portion of the curve represents the proportion that lies outside of plus or minus any particular value of *t*. The column headings indicate these proportions, which can be interpreted as probabilities and are custom-

arily labeled as *α*, the risk of a type 1 error (section 6.3). Note that the proportions of the *t* distribution given in table A.2 and the figure include both "tails" of the distribution. The left column ("df") in table A.2 indicates degrees of freedom. The numbers in the body of the table are values of *t* that correspond to a designated proportion of the distribution for the specified degrees of freedom. Figure 9.1 illustrates a *t* distribution with 4 degrees of freedom. The critical value for probability 0.05 is 2.776, which is found in the table by following the 0.05 column down and the 4 df row across (refer to table A.2 now). Thus, ± 2.776 is the value of *t* that excludes 0.05 of the *t* distribution (or includes 0.95).

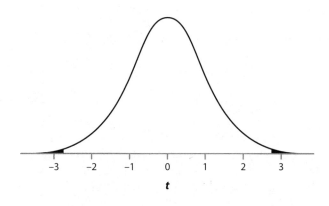

Figure 9.1 A *t* distribution with 4 degrees of freedom. Shaded areas represent 0.05 of the distribution, with 0.025 in each tail.

Suppose we wish to determine the value of *t* that excludes only the upper 0.05 of the *t* distribution with 4 degrees of freedom. Table A.2 gives *t* values that exclude a certain proportion of the distribution equally divided between both tails, as shown in figure 9.1. Thus one-half of the indicated proportion is in the upper tail of the curve, and one-half is in the lower tail of the curve. When we are interested in only one tail, we simply double the proportion indicated in table A.2. For this problem we consult table A.2 for 4 degrees of freedom and a probability (proportion) of 0.10 (rather than 0.05), where we find that 2.132 is the *t* value that excludes the upper 0.05 of the *t* distribution. This is shown graphically in figure 9.2.

Notice in table A.2 that, for any particular proportion (probability), the critical value of *t* becomes smaller as the degrees of freedom becomes larger. At infinite degrees of freedom (effectively, a large number > 120), the value of *t* that excludes 0.05 of the distribution is 1.960—exactly the value of *z* that excludes 0.05 of the standard normal distribution.

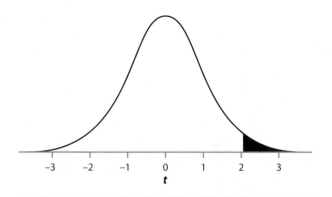

Figure 9.2 A *t* distribution with 4 degrees of freedom. Shaded area represents 0.05 of the distribution in one tail. Compare with figure 9.1.

The reason for this is that as the degrees of freedom increase, the *t* distribution tends to become a normal distribution.

9.3 Estimation: Confidence Interval for μ

We are now ready to construct confidence intervals. Recall that confidence intervals were previously described in sections 6.1 and 8.6. In this case, we would like to generate the confidence interval for the population mean (μ). Using a random sample of size *n*, and the computed values for the sample mean ($\bar{x}$) and sample standard deviation (*s*), we can calculate a range of values within which we have a specified level of confidence that the population mean lies. This range is called a **confidence interval** of the mean. Consider the following example.

Example 9.3

A random sample of 20 male mosquito fish was collected, and total length (in millimeters) was determined. The sample mean length $\bar{x}$ was 21.0 mm, and the sample standard deviation (*s*) was 1.76 mm. We wish to construct a 95% confidence interval for the population mean. Length in male mosquito fish is known to be approximately normally distributed.

Note that both the mean and the standard deviation of the population must be estimated from their sample values, and the sample size is fairly small. The assumptions of this procedure are:

Assumptions of the Test

1. The sample is a random sample from the population of interest.

2. The variable consists of continuous measurements taken from a normal distribution or, if the population is not normally distributed, the sample size is large enough that the means are normally distributed.

The calculations are straight forward. We just need to compute the standard error of the mean ($s_{\bar{x}}$), look up the critical *t* value from table A.2, and put it all together. For a 95% confidence interval, we use α = 0.05 (the column headed with "0.05") and the row corresponding to the appropriate degrees of freedom. For any other confidence interval, just choose the critical *t* for a different value of α (e.g., for 99% CI choose α = 0.01). The tabled value of *t* at α = 0.05 and $n - 1$ degrees of freedom is denoted $t_{(0.05, \, n-1)}$. This critical *t* value delineates 0.95 of the *t* distribution.

The upper limit for a 95% confidence interval is given by

$$UL_{0.95} = \bar{x} + \left(t_{(0.05, \, n-1)} \right) \times s_{\bar{x}} \qquad (9.1)$$

and the lower limit is given by

$$LL_{0.95} = \bar{x} - \left(t_{(0.05, n-1)} \right) \times s_{\bar{x}} \qquad (9.2)$$

where $t_{(0.05, \, n-1)}$ is the tabular value of *t* at $n - 1$ degrees of freedom and $s_{\bar{x}}$ is the sample **standard error of the mean** (section 8.3).

$$s_{\bar{x}} = \frac{s}{\sqrt{n}} \qquad (9.3)$$

Using the statistics reported in example 9.3, the 95% confidence interval is calculated as

$$UL_{0.95} = 21.0 + \left(2.093 \times \frac{1.76}{\sqrt{20}} \right) = 21.825$$

and

$$LL_{0.95} = 21.0 - \left(2.093 \times \frac{1.76}{\sqrt{20}} \right) = 20.175$$

In summary, the 95% CI for μ includes the range {20.175, 21.825}. Thus, we conclude there is a 95% (0.95) probability that the range of 20.175 mm to 21.825 mm includes the population mean.

To illustrate the meaning of a 95% confidence interval, imagine a population with known mean μ which we sample 20 times. For each sample, we compute a 95% confidence interval. We would predict that 19 of the 20 samples (95%) will correctly bracket μ. This situation is illustrated in figure 9.3,

Caution
When thinking about a confidence interval of a mean, or any other parameter, it is incorrect to conclude that there is a 95% (or 99% or whatever) probability that the parameter lies between the upper and lower limits. The parameter has a fixed value that is unknown to us, and therefore there is no probability associated with it: it is whatever it is. The probability in such a case is that our confidence interval includes (or "encloses") the parameter in question.

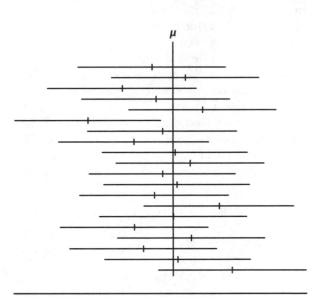

Figure 9.3 A series of twenty 95% confidence intervals (horizontal lines), based on calculations from samples. The position of the true mean, μ, is indicated by the position of the long vertical line.

with confidence intervals from individual samples represented by horizontal lines. Notice that, although the sample means vary around μ, all but one of the confidence intervals brackets μ.

9.4 Reporting a Sample Mean and Its Variation

How should one report a sample mean in a scientific presentation? The mean should always be shown together with some measure of its variation. The only exception is in the case where the entire statistical population is collected and measured. Here the calculated mean is the population mean. But we rarely have that luxury. When the mean is presented without some measure of its variation, the viewer is cheated out of some key information.

There are several common methods for presenting these key statistics, depending on the information one wishes to convey. The information in example 9.3 might be reported in any one of the following ways.

1. The mean plus or minus the standard deviation $(\bar{x} \pm s)$, which for example 9.3 is 21.0 ± 1.76. While this tells the reader something about the variation of the measured variable in the population, it provides little information about how well the sample mean estimates the population mean, unless the reader wishes to calculate a confidence interval for himself or herself.

2. The mean plus or minus the standard error $(\bar{x} \pm s_{\bar{x}})$. This is perhaps the most common way of reporting a sample mean, and it does provide a rough idea of how well the sample mean estimates the population mean. However, it does not give this information as precisely as does a confidence interval.

3. The mean plus or minus the 95% (or 99%) confidence interval, or $\bar{x} \pm CI_{(0.95)}$. This method of reporting a sample mean has a great deal to recommend it. The information conveyed is directly accessible to the reader and is not misleading. Nevertheless, providing the standard error and the sample size allows the reader to determine the CI themselves.

Sample means are often represented graphically, using one of these three measures of variation. The most common form of such a graph is shown in figure 9.4. The points on the graph represent sample means, and the vertical lines (called "error bars") represent plus or minus one standard error of the mean.

As a general rule, *always plot your data*, even if it's only for your own inspection. We can add to that another general rule: *always show the variation in the data*, as well as the average. This second rule applies to summary tables as well as figures.

9.5 Hypothesis Concerning a Single Population Mean

We are now prepared to consider testing a hypothesis about a single population mean. Before going further, refer back to chapter 6 to review the theory behind hypothesis testing. In particular, you should be familiar with statistical hypotheses and the interpretation of p-values.

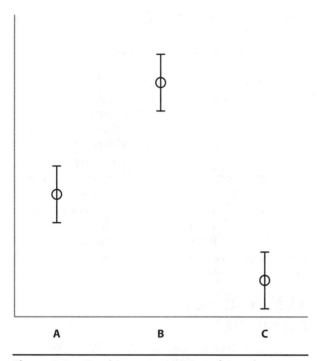

Figure 9.4 Graphic representation of sample means and their standard errors.

9.5.1 One-Sample *t* Test

In general, a one-sample *t* test is used to evaluate the following null hypothesis

$$H_0: \mu = \mu_0$$

where μ_0 is some specific number. Consider example 9.1 above. We are interested in testing whether or not the population mean (μ) for vitamin X content is 100 units (the specified FDA requirement). A sample of 50 pills had a mean of 100.5 and sample standard deviation of 2.19 units per pill.

In this situation, the null hypothesis to be tested is that the mean vitamin X content of the pills is 100 units, or

$$H_0: \mu = 100 \text{ units}$$

Note the equal sign in this expression. The alternative hypothesis (the one that is true if the null hypothesis is false) is

$$H_a: \mu \neq 100 \text{ units}$$

We assume, for the moment, that the population mean of these vitamin pills is 100 units. We do not know if this is true; it is the population mean specified by the null hypothesis, and we wish to determine if it might be true. If we conclude that it is not true, then we conclude that the only other possibility—the alternative hypothesis—is true.

Recall from the discussion of sampling distributions of means (section 8.3) that there is a certain probability that a sample with a mean of 100.5 units could be drawn from a population with a population mean of 100 units by chance. Remember, we do not know the population mean. However *if* the population mean were 100, *then* there is a certain probability associated with drawing a sample with a mean that differs by as much as 0.5 units from the population mean of 100. To test the null hypothesis, we calculate a *t* statistic (t_s) by the following formula:

$$t_s = \frac{\bar{x} - \mu_0}{s/\sqrt{n}} \qquad (9.4)$$

where μ_0 represents our hypothesized value (in this example, 100), and $s/\sqrt{n}$ we recognize as the formula for the standard error of the mean (eq. 9.3). When the *t* statistic is equal to or greater than the critical value of *t* for a specified area of the *t* distribution (usually 0.05), we reject the null hypothesis, since the probability of obtaining such a value of *t* when the null hypothesis is true is quite small (0.05 or less). The critical value of *t* is obtained from table A.2.

The assumptions of the *t* test are the same as those for computing a confidence interval using the *t* distribution (section 9.3).

For the example, with $n = 50$, the critical value of *t* for $\alpha = 0.05$ and 49 degrees of freedom is ca. 2.011. Note that, since the exact value for 49 degrees of freedom was not shown, we had to **interpolate**—i.e., calculate an average between the two tabled values (2.021 and 2.000).

This problem is shown graphically in figure 9.5 (top) on the following page. The *x* axis represents the distribution of *t* that would be expected under the null hypothesis. In other words, if the null hypothesis were true, then the values +2.011 and –2.011 delimit 0.95 of the distribution. Thus, there is a probability of 0.95 that any sample we might take from this population would have a *t* value of between –2.011 and +2.011, and a probability of 0.05 that any sample would have a value lower than –2.011 or greater than +2.011. The shaded areas—the proportion of the curve beyond ±2.011—are sometimes referred to as the **rejection region** because when the *t* statistic falls within this region, the null hypothesis would be rejected. In other words, if such a statistic were highly unlikely, then such a low probability would be grounds for rejecting the null hypothesis.

Let's complete our example. Using the reported statistics and equation 9.4,

$$t_s = \frac{100.5 - 100}{2.19/\sqrt{50}} = 1.614$$

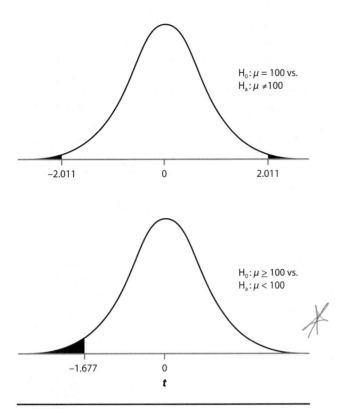

Figure 9.5 Comparison of *t* distributions at 49 degrees of freedom and $\alpha = 0.05$ for two-tailed (top) and one-tailed (bottom) hypotheses. The rejection region in the top figure has 0.025 in each tail, for the bottom figure 0.05 in one tail.

Since the *t* statistic (1.614) does not exceed the critical value (2.011), we *fail to reject* H_0. (Notice that we have not proven that H_0 is true; the best we can say is that the data are consistent with the null hypothesis.)

Consulting table A.2 further, we can determine the p-value. Recall that the **p-value** is the probability of obtaining a statistic as large or larger, if the null hypothesis were true. At the appropriate degrees of freedom (use rows for 40 and 60), our *t* statistic falls between the tabled critical values for probabilities of 0.1 and 0.2. We thus report our p-value as: $0.1 < p < 0.2$. From this example, we conclude that the mean vitamin X content of the pills is no different than 100 units. The fact that *p* falls between 0.1 and 0.2 indicates that when the null hypothesis is true that such samples should occur

---

Caution

One way of evaluating any null hypothesis is to compare the p-value to α. We reject the null hypothesis only when the p-value is less than α. One should always report the p-value, which allows another person to come to their own interpretation of the data.

---

between 10% and 20% of the time—i.e., this is pretty likely.

We should bear in mind that it's possible that this decision is mistaken (a type 2 error). Refer back to section 6.3 for a review of errors in hypothesis testing. Nevertheless, in the current example, the sample size is large (50) and the sample mean (100.5) is quite close to the hypothesized value. Thus, we should be comfortable in supporting the null hypothesis.

To finish the example, we can compute a confidence interval for the vitamin X content. Refer back to the earlier section for the formula and compute that confidence interval now. You should get an interval of 100.5 ± 0.623 or {99.9, 101.1}. Notice that the confidence interval includes the hypothesized value of 100. This further confirms our interpretation of the results.

9.5.2 One-Tailed and Two-Tailed Hypothesis Tests

Note that in the above example we have included one-half of alpha, ($\alpha/2 = 0.025$) in each tail of the distribution (fig. 9.5 top). This is because the null hypothesis specified that $\mu = 100$, and if rejected, it would tell us only that $\mu \neq 100$. Thus, we would reject H_0 if *t* fell in either the upper or the lower tail of the distribution (i.e., $t \leq -2.011$, or $t \geq 2.011$). This is called a **two-tailed test** (also called a two-sided test). In two-tailed tests, it is the absolute value of *t* that is important.

In some situations we have an interest in only one tail of a distribution, and these are referred to as **one-tailed tests**. This can be illustrated by changing our example slightly. Suppose the FDA requirement for vitamin pills specifies that pills must contain an average of *at least* 100 units of vitamin X per pill. They may contain 100 units or more, but they may not contain less than 100 units. The null hypothesis for this case is

$$H_0: \mu \geq 100$$

and the alternative hypothesis is

$$H_a: \mu < 100$$

A frequent problem for students is in deciding which should be the null hypothesis and which should be the alternative hypothesis. One guideline is that, since it must generate a single probability distribution, the null hypothesis must contain an equal sign. In the example, $\mu = 100$ allows generating a single probability distribution. Another guideline is that the null hypothesis is the conservative position (or "straw man"), which must be disproven "beyond a reasonable doubt." Refer back to section 6.2 for further examples. In this example, the conservative position is that the FDA guidelines are met. Only if the vitamin X content is significantly lower than 100 will the manufacturer fail to meet the guidelines.

In this situation our interest is only in the lower tail of the t distribution. The shaded portion of figure 9.5 (bottom) represents 0.05 of the t distribution when our interest is only in the lower tail.

The value of t that delimits this area of the t distribution is -1.677. This critical value is found in table A.2 by looking in the column headed "0.1" at 100 degrees of freedom. (Table A.2 gives only two-tailed probabilities. To find a one-tailed probability, we must consult the column for twice the desired probability.) A portion of the t distribution is copied below in table 9.1 to show the relationship between critical values for two-tailed and one-tailed tests.

In the current example, with 49 df, a t_s value of -1.677 or lower would be justification for rejecting the null hypothesis. Since our calculated value for t_s is much higher than this (a *positive* 1.614), we cannot reject H_0. We conclude that the vitamin pills contain 100 or more units of X per pill. Note that this one-tailed test does not permit us to conclude that the pills contain exactly 100 units or more than 100 units; it only indicates that they do not contain less than 100 units. Note also that in a one-tailed test, the sign of t is important.

The same procedure would be used if we wished to conduct a one-tailed test with alpha only in the upper tail of the distribution. In this case the null hypothesis would be that the pills contain no more than 100 units, or H_0: $\mu \le 100$, and the alternative hypothesis is H_a: $\mu > 100$. In this case the critical value of t is $+1.677$, and we would reject H_0 when our calculated value of t is equal to or greater than this value. For the example, our calculated value of $t_s(1.614)$ is not greater than 1.677. So we would again fail to reject H_0 and conclude that the pills contain no more than 100 units. This situation is shown in figure 9.6.

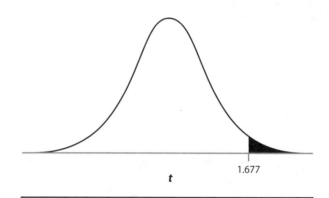

Figure 9.6 A t distribution with 49 degrees of freedom for an upper one-tailed probability, $\alpha = 0.05$.

How does one know when a one-tailed test is appropriate and when a two-tailed test is appropriate? That depends on the null hypothesis, which in turn depends on the question we wish to answer. The three questions we asked concerning the vitamin pill example are fairly typical (although vitamin X is an imaginary nutrient). If we wish to know if the vitamin content of the pills is either more than or less than 100 units, a two-tailed test is indicated. On the other hand, if interest centers only in determining if the content is more than 100 units, we would choose a one-tailed test. Thus, it is very important to state the null hypothesis in such a way that, if it is rejected, we will arrive at a sensible answer to the research question we had in mind.

We will consider another example of a one-tailed hypothesis test in section 9.6 below.

Table 9.1 Critical Values of the t Distribution. A portion of table A.2 was copied (two-tailed portion shown in top half) to show how one-tailed values can be determined (in the lower half). The probabilities may represent either α or p-values, depending on how the table is used.

Two-tailed

| df | \multicolumn{7}{c}{Probability} |
	0.2	0.1	0.05	0.02	0.01	0.001	0.0001
15	1.341	1.753	2.131	2.602	2.947	4.073	5.239
16	1.337	1.746	2.120	2.583	2.921	4.015	5.134
17	1.333	1.740	2.110	2.567	2.898	3.965	5.044
18	1.33	1.734	2.101	2.552	2.878	3.922	4.966
19	1.328	1.729	2.093	2.539	2.861	3.883	4.897
20	1.325	1.725	2.086	2.528	2.845	3.850	4.837

One-tailed

| df | \multicolumn{7}{c}{Probability} |
	0.1	0.05	0.025	0.01	0.005	0.0005	0.00005
15	1.341	1.753	2.131	2.602	2.947	4.073	5.239
16	1.337	1.746	2.120	2.583	2.921	4.015	5.134
17	1.333	1.740	2.110	2.567	2.898	3.965	5.044
18	1.33	1.734	2.101	2.552	2.878	3.922	4.966
19	1.328	1.729	2.093	2.539	2.861	3.883	4.897
20	1.325	1.725	2.086	2.528	2.845	3.850	4.837

Caution

Choice of a one-tailed or two-tailed hypothesis must come from the particular research question we have in mind. It is never appropriate to first examine the data and then decide which hypothesis test is appropriate.

9.5.3 Nonparametric Alternative to the *t* Test: Wilcoxon Signed Rank Test

Often, researchers would like to evaluate a hypothesis about a single population mean, but the particular data do not meet the assumptions of the *t* test. Assumption 2 above states: "The variable consists of continuous measurements taken from a normal distribution or, if the population is not normally distributed, the sample size is large enough that the means are normally distributed." If this assumption fails, then we can still evaluate the hypothesis, but in a different way.

T tests are examples of the broad category of **parametric tests**. Parametric tests depend on an underlying normal distribution. Many other statistical tests covered in this book are also parametric tests (e.g., ANOVA, regression, correlation). When the assumptions of parametric tests are not met, one way around this road block is to run a **nonparametric test**. Nonparametric tests are sometimes called "distribution-free tests," but this statement is not entirely accurate. It's better to say that nonparametric tests do not depend on a normal distribution. Nonparametric tests are available as analogs to many of the parametric tests and are often used when underlying distributions are skewed and sample sizes are too small for the central limit theorem (chapter 8) to save us. Nonparametric tests can also be used for data that are not measurements (e.g., ranks).

The nonparametric analog to the one-sample *t* test is called the Wilcoxon signed rank test. In place of the original null hypothesis about the mean being equal to some specific value (H$_0$: $\mu = \mu_0$), the Wilcoxon test evaluates a null hypothesis about the median (H$_0$: $\theta = \theta_0$). The Wilcoxon test is explained more fully with a matched-pairs example below (section 9.6.2). When the Wilcoxon test is used with a single sample, subtract θ_0 from each measurement and then follow the formula shown in section 9.6.2.

9.6 Tests for Two Related Samples: Think Differences

One of the most common uses of hypothesis testing is to determine if a particular treatment has an effect on some response variable of interest. Does dieting affect cholesterol levels in the blood? Do two species of snakes exhibit the same feeding rates on rodent prey? There are two basic approaches for conducting these experiments: collecting two independent samples or matching the treatments in some way, such as with a before and after study.

Comparisons of means from two independent samples are discussed in chapter 10. In the following section we explore the statistics with experiments or surveys that employ **matched pairs** (also called **paired-comparisons**). In this design, each experimental subject is matched with another subject that is as similar as possible and multiple pairs are randomly chosen. One member of each pair receives one treatment and the other member receives the second treatment. For example, we might compare the growth responses of plants to two different fertilizer treatments, where each pair of plants is on a different greenhouse bench. Another example would be when each human subject has cholesterol samples drawn both before and after a period of dieting. This experimental design is also called **blocking**, a method that can be applied to any number of treatments (sections 12.3 and 15.1). An advantage of blocking is that it reduces the amount of unexplained variation among subjects that has nothing to do with the treatment of interest.

9.6.1 The Paired *t* Test

When measurements are continuous and when the variable under study may be assumed to be approximately normally distributed, the *t* test for paired (matched) samples may be used. This test is equivalent to the one sample *t* test discussed above since, in effect, we are dealing with one sample—the difference in the observed individuals before and after some treatment, or the difference between matched pairs of individuals under different specified conditions.

Assumptions of the Test

1. Measurement is continuous or, if discrete, the range of possible values is large.

2. The distribution of the variable (differences) is approximately normal.

3. Each individual is measured twice, once before the specified treatment and again following the specified treatment; or matched pairs of individuals are measured.

4. The data constitute a random sample from the population of interest.

As an example of a paired *t* test, consider a study of burn healing (example 9.2 above). Here we are interested in comparing new (N) versus old (O) medications in their effect on speed of recovery from skin burns. The test involved 20 patients who had severe burns on both arms and used a matched-pairs design, with medication N on one arm and O on the other arm and random selection of arms to treatment. To avoid bias, the clinician who diagnosed the healing of the arms was not told which treatment each arm had received. Following the course of treatment, the time (in days) necessary for wound healing was measured on each arm. Since the scientist conducting the study expected the healing rate to be faster (hence shorter duration) for the new medication than for the old medication, she computed the difference D = O – N to represent improvement of the new medication to the old. The data follow. We would like to know if these data support the hypothesis that the new medication results in faster wound healing than the old medication.

Table 9.2 Wound Healing Time in Patients Receiving Old and New Medications on Different Arms

| Patient | Wound Healing Time (Days) | | |
	Medication Old (O)	Medication New (N)	Difference (O – N)
1	23.5	19.5	4.0
2	22.3	20.1	2.2
3	20.2	19.1	1.1
4	22.5	19.2	3.3
5	21.3	21.2	0.1
6	23.2	19.6	3.6
7	23.7	19.8	3.9
8	22.7	20.2	2.5
9	20.2	17.7	2.5
10	20.9	18.6	2.3
11	22.3	20.8	1.5
12	23.1	21.1	2.0
13	20.3	18.5	1.8
14	25.1	16.1	9.0
15	24.8	17.3	7.5
16	22.0	22.4	-0.4
17	21.3	21.5	-0.2
18	23.2	18.6	4.6
19	21.3	20.5	0.8
20	22.0	20.4	1.6

Before we go further, let's consider our hypotheses. The working hypothesis is that wound healing time is shorter in the arm receiving the new medication than the arm receiving the old medication. Thus we expect that on average, the differences (D = O – N) will be positive. This is clearly a one-sided hypothesis (similar to that depicted in fig. 9.6, but with a different t critical value). The null hypothesis for this one-tailed test is that the population mean difference (μ_D) in healing time between old and new treatment is less than or equal to zero, or

$$H_0 : \mu_D \leq 0$$

so there is no difference btw the two

and the alternative hypothesis would be

$$H_a : \mu_D > 0 \quad \text{there is a difference}$$

Recall that the critical value of t is for a one-tailed test when alpha is set at 0.05 is found in table A.2 in the column headed "0.1." With a sample size of 20, df = 20 – 1 = 19. The t critical value is 1.729. We will reject the null hypothesis if the t statistic exceeds this critical value.

Using the data in table 9.2, the mean of these differences ($\bar{x}_D$) is 2.685 days, with a standard deviation of 2.362 days. The expression

$$t = \frac{\bar{x}_D}{s_{\bar{D}}} \quad \text{standard error of the mean diff} \quad (9.5)$$

has a t distribution with n – 1 degrees of freedom, where n is the number of matched pairs. The symbol $\bar{x}_D$ is the sample mean difference, and $s_{\bar{D}}$ is the standard error of the mean difference, calculated by

$$s_{\bar{D}} = \frac{s_D}{\sqrt{n}} \quad \frac{2.362}{\sqrt{20}} = \quad (9.6)$$

[Note the distinction between the standard deviation (S_D) and the standard error ($s_{\bar{D}}$)]. For the example

$$s_{\bar{D}} = \frac{s}{\sqrt{n}} = \frac{2.362}{\sqrt{20}} = 0.528$$

and

$$t = \frac{\bar{x}_D}{s_{\bar{D}}} = \frac{2.685}{0.528} = 5.08$$

Clearly, the t statistic exceeds the critical value. The p-value, determined by consulting table A.2, is $p < 0.00005$ ($p < 0.0001 \div 2$). Therefore, we reject H_0 and conclude that healing time is faster with the new medication than with the old medication.

Based on the difference in healing times relative to the time for the old medication, the new medication offers an average of 12% faster healing than the old medication. This seems not only statistically significant but also medically significant as well.

If we wish, we could compute a confidence interval for the mean difference (μ_D) by the same method used earlier for a single sample (section 9.3). Doing that, you should find that the 95% confidence interval for the difference in healing time in days is {1.58, 3.79}.[2] In other words, with 95% confidence, we can say that the mean difference in healing times is at least 1.58 and as much as 3.79 days.

9.6.2 Nonparametric Tests for Two Related Samples

In matched pairs or repeated measures experiments in which the data do not satisfy the assumptions of the parametric test, one of two nonparametric tests for two related samples may be used: the Wilcoxon signed rank test and the sign test. The Wilcoxon signed rank test is the more commonly-used test, so we consider that first.

The Wilcoxon Signed Rank Test

This nonparametric test is appropriate when the direction of the difference between matched pairs can be determined and when the differences can be ranked with respect to each other. Thus, it is appropriate for data that consist of ranks (i.e., are not measurements) or for measurement variables which are not normally distributed. This test gives greater weight to pairs with larger differences than to pairs with smaller differences. The differences are ranked with respect to their absolute values (i.e., –1 has a lower value than either +2 or –2), but the sign of the difference is retained with the rank. Pairs in which there is no difference are dropped from the sample, and the sample size is reduced accordingly.

For this test, when H_0 is true, the sum of the positive ranks in a population should be about equal to the sum of the negative ranks in the population.

Note that animals 3 and 7 were tied with a rank of 1.5. The fact that animal 3 had a difference in before and after scores of –1 and animal 7 had a difference of +1 makes no difference with respect to their ranks. However, the fact that animal 3 had a negative score is noted. To calculate the value of the test statistic, designated as T, the sum of ranks that have a negative sign and the sum of ranks that have a positive sign are taken. T is the smaller of these two sums. For the example, T is 1.5. Critical values of T are given in table A.5. H_0 is rejected if the calculated value of T is equal to or *smaller* than the tabular value of T. In this case, the critical value is 2, so we reject the null hypothesis and conclude that gully cats are more aggressive when kittens are present.

Statistical packages such as *Minitab* readily run nonparametric tests. Results for example 9.4 are given in table 9.3.

The Wilcoxon test was run on the differences in aggressiveness scores between gully cats when they had kittens and when they did not have kittens. The difference in score was on average (median) 3.75 higher when the gully cats had kittens than when they did not have kittens. Notice that the Wilcoxon test was run to test the one-sided hypothesis. The statistic was computed differently than the approach we used above, but arrived at the same conclusion. The low p-value ($p = 0.021$) is less than our typical value of α (0.05), and so we have grounds to reject the null hypothesis and conclude that female gully cats are more aggressive when they have kittens than when they are without kittens.

Table A.5 gives values of n (sample size) up to 25. For larger samples, T is approximately normally distributed and the normal distribution may be used, where

$$z = \frac{T - \frac{n(n+1)}{4}}{\sqrt{\frac{n(n+1)(2n+1)}{24}}} \qquad (9.7)$$

For hypothesis tests at $\alpha = 0.05$, H_0 is rejected if |z| is equal to or larger than 1.96.

Example 9.4
The Wilcoxon Matched-Pairs Test

Female gully cats are thought to be more aggressive when they have kittens. Accordingly, aggressiveness scores on a scale of 1–10, with 10 being most aggressive, were obtained for a group of 7 females without kittens and for these same 7 females when they had kittens. The scores were as follows.

Female ID #	Without Kittens	With Kittens	D	Rank
1	3	7	4	4
2	2	8	6	6
3	5	4	–1	1.5(–)
4	6	9	3	3
5	5	10	5	5
6	1	9	8	7
7	8	9	1	1.5

Table 9.3 *Minitab* Results for Example 9.4

Wilcoxon Signed Rank Test: Dif (with-without)

Test of median = 0.000000 versus median > 0.000000

	N	N for Test	Wilcoxon Statistic	P	Estimated Median
Dif (with-without)	7	7	26.5	0.021	3.750

The Sign Test

The sign test is a nonparametric test for paired or matched samples that is useful when the direction (< or >) of the difference between matched pairs can be determined but when the magnitude of the difference cannot. The null hypothesis tested is that $p(A > B) = p(A < B)$, where A and B are measurements of the matched pair. In plain English, this means that the probability that A is greater than B is equal to the probability that A is smaller than B for any given pair.

To conduct the test, each pair is given either a plus or a minus sign, depending on which member of the pair is larger. Tied pairs (no difference) are dropped from the analysis, and n is reduced accordingly. This is essentially a binomial distribution problem in which we compare the frequency of pluses and minuses with the distribution expected under the null hypothesis:

H_0: frequency of pluses = frequency of minuses

with $p = 0.5$, $q = 0.5$, and k = the number of paired observations (n). This is exactly analogous to determining the probability of obtaining x heads and $k - x$ tails in k tosses of a coin (section 5.4).

Example 9.5
The Sign Test

A football team during summer training was given a new sports beverage. Fifteen minutes after drinking it, each player was asked if he felt better than, worse than, or the same as before he drank the beverage. 9 reported feeling better, 1 reported feeling worse, and 2 reported no change. Using this sample of 12 players, we wish to know if drinking this beverage affects how players feel.

In this situation each individual is paired with himself or herself and is in effect measured twice—feeling before the beverage and feeling after the beverage. Thus, we can detect the direction of change in any individual (better or worse), but not the magnitude of the change. The sign test is appropriate for such a situation, and it is conducted in the following manner.

Nine players reported feeling better and would receive a plus sign, while 1 reported feeling worse and would receive a minus sign. The two who reported no change are dropped from the sample, and the sample size is reduced accordingly. Let x = the frequency of the less frequent sign (1, in this case) and $k - x$ = the frequency of the more frequent sign (9, in this case). The probability of obtaining 1 minus

sign and 9 plus signs in 10 "trials" ($k = 10$) is given by the now familiar expression (from section 5.4):

$$p(x) = \frac{k!}{x!(k-x)!} p^x q^{(k-x)} \qquad (9.8)$$

As before, we are interested in the probability of the exact outcome we observed plus any even more extreme outcomes (in this case 0 minuses and 10 pluses is the only more extreme outcome). For the example

$$p(1) = \frac{10!}{1! \times 9!} 0.5^1 \times 0.5^9 = 0.009760$$

and

$$p(0) = \frac{10!}{0! \times 10!} 0.5^0 \times 0.5^{10} = 0.000976$$

$$p(1) + p(0) = 0.009876 + 0.000976 = 0.010736$$

or a probability of approximately 0.011 of obtaining 1 minus value and 9 plus values or an even more extreme case if the null hypothesis is true. This is a one-tailed probability, but the way the example is phrased ("better" or "worse") indicates a two-tailed test. Thus, the probability of obtaining the observed result or one even more extreme if H_0 is true is twice the one-tailed probability, or $0.010736 \times 2 = 0.021472$. Since the p-value is less than 0.05 (our usual choice for α), we reject H_0 and conclude that drinking the beverage affected how the players felt.

Why is this a two-tailed test? We do so because we did not specify the direction of change in feeling in the null hypothesis. Had our research hypothesis been that taking the new sports drink makes players feel better, the null hypothesis would have been that taking the drink makes players feel no different or worse. Had this been the null hypothesis, it could be rejected with a p-value of 0.010736 rather than one of 0.021472.

Summary

The current chapter examined tests of hypotheses about single population means. A single random sample is collected, either directly from some population or indirectly, following matching of treatments in some paired experimental design. If the data follow the assumptions of parametric statistics, the one-sample t test can be applied to the sample data (or differences for the matched-pairs study). If the data do not follow the parametric assumptions, then the usual procedure is to conduct one of the analogous nonparametric procedures. In the next chapter, we will explore hypothesis tests for comparing two statistical populations after collecting two independent random samples.

Key Terms

blocking
confidence interval
degrees of freedom (df)
interpolate
matched pairs
nonparametric
one-tailed test
p-value
paired comparison
parametric
rejection region
sign test
t distribution
t test
two-tailed test
Wilcoxon signed rank test

$H_0 = \mu = 100$
$H_1 = \mu \neq 100$

$n = 20$
$\bar{x} = 97$
$S = 2.5$
$\mu_0 = 100$

$\bar{x} = 267$
$S = 12.1$

$H_0 = \mu = 250$
$H_1 \quad \mu \neq 250$

Exercises

For all hypothesis tests, be sure to specify the null (H_0) and alternative hypotheses, conditions for rejecting the H_0 (at $\alpha = 0.05$), your resulting test statistic, decision about validity of H_0, p-value, and verbal conclusion. First work the problems by hand. If a dataset is provided, solve both by hand and with statistical software and compare your answers.

CONFIDENCE INTERVALS

9.1 The pulse rate (in bpm) of a random sample of 30 Peruvian Indians was collected. The mean pulse rate of the sample is 70.2, with a sample standard deviation of 10.51. Compute a 95% confidence interval for the population mean. (Data adapted from Hand et al. 1994)

$n = 30$
$\bar{x} = 70.2$
$S = 10.51$

9.2 The activity of o-diphenol oxidase ($\mu l O_2/mgP/min$) was measured in 15 tomato plants. The mean activity in the sample was 35.4, with a standard deviation of 6.27. Compute the 95% confidence interval for the population mean.

$n = 15$
$\bar{x} = 35.4$
$S = 6.27$

9.3 The weights (in milligrams) of spleens of 9 newly hatched turkeys are given below. Compute the 95% and the 99% confidence intervals for the population mean. (Data from F. McCorkle)

$n = 9$
$\bar{x} = 19.25$
$S = 2.56$

18.9	20.4	15.9
19.9	17.4	24.0
21.3	16.2	19.3

9.4 The snout vent lengths (in centimeters) in 25 newly born garter snakes selected at random from several litters are given below. Compute

the 95% and the 99% confidence intervals for the population mean.

6.5	4.3	4.6	6.0	4.7
6.2	5.8	5.4	5.2	4.8
4.9	5.0	4.7	3.4	3.9
5.1	5.4	4.8	3.8	6.1

ONE-SAMPLE t TEST

9.5 A turkey geneticist wished to select for breeding purposes hens whose eggs have an average weight of 100 grams. Hens with eggs less than or greater than this mean weight are unacceptable. The mean weight of a sample of 20 eggs from one hen was 97 grams, with a standard deviation of 2.5 grams. Should this hen be kept in the breeding program?

9.6 The label on a company's energy drinks claims that they contain an average caffeine concentration of 250 mg/oz. The mean caffeine concentration of 15 randomly sampled drinks was 267, with a standard deviation of 12.1. Are the drink labels accurate?

9.7 Suppose that the manufacturer of UltraChic cigarettes claims that the average nicotine content does not exceed 3.5 mg per cigarette. Based on prior studies, we know that nicotine content tends to be normally distributed. We collect a random sample of 10 cigarettes, make measurements in an independent laboratory, and determine the sample mean to be 4.2 mg, with a standard deviation of 1.4 mg nicotine per cigarette. Are these data consistent with the manufacturer's claim?

9.8 The FDA has established that the concentration of a certain pesticide in apples may not exceed 10 ppb. A random sample of 100 apples from a major orchard had an average pesticide content of 10.03 ppb with a standard deviation of 0.12 ppb. Are the apples within the FDA requirement?

9.9 A certain enzyme in the liver of fish is considered an indicator of trace amounts of a dangerous pollutant that is difficult to detect by chemical methods. Enzyme activities of less than 50 units per gram of liver (fresh weight) are taken to indicate the presence of the pollutant. A random sample was taken from a local stream and the enzyme concentrations were as follows. Are enzyme activities depressed in this population?

48	43	51	42
50	42	44	45
56	49	44	47
50	49	38	46
38	52	32	56

$n = 20$
mean =
stdv =
$\mu_0 = 50$

9.10 The FDA sets a standard for no more than 1.0 ppm of mercury in commercially caught fish. The mean concentration of mercury in 15 randomly sampled big-eye tuna was 1.086 ppm, with a standard deviation of 0.114 ppm. Does the sample fall within the standards for the FDA?

1.030	1.129	1.262	1.159	1.112
1.113	1.057	1.231	0.869	1.231
1.052	0.884	1.039	1.024	1.105

9.11 Healthy populations of mythical elephant birds have newborn chicks that weigh, on average, 8.0 kg. We expect that chicks from populations impacted by disease to weigh less. Suppose we visit a new population and sample 20 randomly selected chicks. Based on the measurements below, test whether or not this population has chicks that are below normal in weight.

8.1 8.0 8.0 7.7 8.1 7.9 7.6 7.6 7.9 8.1
7.8 8.2 8.1 8.3 8.0 8.0 7.8 7.9 7.6 8.1

9.12 Suppose that the mean water hardness of lakes in Kansas is 425 mg/L and these values tend to follow a normal distribution. A limnologist would like to know whether stock ponds tend to have lower hardness. He collected water from 25 randomly-selected stock ponds, which yielded the following results. Test the appropriate null hypothesis.

346 496 352 378 315 420 485 446 479 422
494 289 436 516 615 491 360 385 500 558
381 303 434 562 496

THE PAIRED *t* TEST

9.13 Data on resting and post-exercise pulse rates were collected for 8 individuals between 19 and 22 years of age. We wish to know if there is a difference in pre-exercise and post-exercise pulse rates.

Individual Number	Resting	Post-exercise
1	108	136
2	60	90
3	70	78
4	54	108
5	54	102
6	72	92
7	101	118
8	96	176

9.14 For the situation in 9.13, the following data are for pre-exercise and post-exercise body temperature. Is there a difference in the two body temperatures?

Individual Number	Resting	Post-exercise
1	99.0	99.4
2	97.8	98.1
3	98.6	98.6
4	98.7	98.7
5	98.7	98.7
6	98.2	98.2
7	98.7	98.8
8	98.6	99.2

9.15 Long-term exposure to stress can result in a variety of health problems. A study tested the effects of cortisone, the primary hormone released during stress, on memory retention of 12 individuals. Subjects were shown a set of 60 words and 24 hours later were asked to recall as many words as they could remember. One hour before the retention test subjects received a dose of either cortisone or a placebo. Then two weeks later the same procedures were repeated with the same individuals, but with a different set of 60 words and with subjects receiving the opposite treatment than they received the first week. The number of words retained for both treatments is listed below. Does cortisone decrease long term memory?

Individual Number	Placebo	Cortisone
1	14	11
2	16	12
3	11	10
4	12	9
5	15	13
6	15	12
7	17	14
8	14	9
9	15	11
10	12	9
11	16	10
12	13	14

9.16 The wattle thickness (in millimeters) of 15 randomly selected chickens was measured before and after treatment with PHA. Does treatment with PHA increase wattle thickness? (Data from F. McCorkle)

Chicken Number	Pretreatment	Post-treatment
1	1.05	3.48
2	1.01	5.02
3	0.78	5.37
4	0.98	5.45
5	0.81	5.37
6	0.95	3.92
7	1.00	6.54
8	0.83	3.42
9	0.78	3.72
10	1.05	3.25
11	1.04	3.66
12	1.03	3.12
13	0.95	4.22
14	1.46	2.53
15	0.78	4.39

THE WILCOXON SIGNED RANK MATCHED-PAIRS TEST

9.17 Evolutionary theory and empirical evidence suggest that an animal infected with a parasite will change their social behavior in order to increase transmission of the parasite. A study was conducted to test whether human social interactions increase in response to the common flu vaccine (a proxy for natural infection). The number of people encountered 48 hours before immunization was compared to the number of people encountered 48 hours after immunization in 12 individuals. Does the vaccine increase the encounter rate? Prior studies indicated that similar data fail the assumptions for parametric tests.

Individual ID	Before Immunization	After Immunization
1	90	94
2	59	100
3	56	85
4	72	84
5	66	106
6	56	91
7	61	60
8	44	112
9	58	127
10	68	67
11	66	125
12	81	108

9.18 Ten individuals were asked to rate their feeling of well-being on a rank scale of 1 to 10 before and after taking an experimental drug. Does the drug change a person's sense of well-being?

Individual Number	Before Drug	After Drug
1	5	7
2	8	9
3	2	1
4	7	9
5	5	5
6	2	9
7	9	9
8	3	9
9	9	10
10	6	7

THE SIGN TEST

9.19 Ten subjects were given an experimental drug and asked if their sense of well-being improved, became worse, or did not change after taking the drug. We wish to know if the drug is effective in improving one's sense of well-being. The results were as follows:

7 reported an improvement
2 reported no change
1 reported feeling worse

9.20 A chocolate company wishes to know if its consumers prefer dark chocolate or milk chocolate more. They survey 16 individuals with the following results:

3 preferred milk chocolate
5 no preference
8 preferred dark chocolate

Notes

[1] "Student" was the pseudonym for William Gossett, a statistician employed by Guinness Brewery in the early 1900s.

[2] This is a two-sided confidence interval. We have not presented the method for computing a one-sided interval, although statistical packages such as *Minitab* will do this automatically, as part of a one-sided *t* test.

chapter ten

Inferences Concerning Two Population Means

One of the more commonly used groups of statistical tests are those designed to test whether two or more populations differ from each other in some way. How does resting heart rate differ between males and females? Between smokers and non-smokers? In both these cases, we would survey different populations. How does treatment with copper sulfate affect the density of mucus cells on the gills of bluegill sunfish? Does a new type of inhaler provide more relief to people suffering from asthma than the old type of inhaler? In these two cases we could take a large group of individuals (sunfish or people with asthma) and randomly assign samples to two different treatments. We are most often interested in differences in the means of these populations, but we can also test for differences in variability or distribution.

The current chapter considers methods for dealing with questions about whether two population means differ. This hypothesis requires that two **independent samples** are randomly collected from the populations of interest. Contrast this design with **related samples**, or as they are often called, matched-pairs samples (section 9.6). In related-samples experiments, each individual is measured twice or carefully matched pairs of individuals are measured.

When its assumptions are met, the most powerful test for comparing two independent populations is the two-sample t test. The t test is an example of a parametric test. A very useful nonparametric test that can usually be used when the data do not meet the assumptions is the Mann-Whitney U test. Both tests are discussed in the following sections.

10.1 The t Test for Two Independent Samples

Recall from the previous discussion of sampling distributions (chapter 8) that the means of samples from a population of unknown variance follow a t distribution if the underlying distribution for the measured variable is normal or approximately normal. We used this concept in chapter 9 to make inferences about a single population mean. By a similar line of reasoning, we may use the t distribution to test whether two population means differ from each other. The null hypothesis of such a test is that the means of the two populations, a and b are equal, or

$$H_0: \mu_a = \mu_b$$

If repeated sample means from a population have a t distribution, then it follows that the difference between repeated *pairs* of sample means, taken from a population or from two populations with the same population mean, also have a t distribution.

$$t = \frac{(\bar{x}_a - \bar{x}_b) - (\mu_a - \mu_b)}{s_p} \quad (10.1)$$

where $(\bar{x}_a - \bar{x}_b)$ is the difference between the two sample means and $(\mu_a - \mu_b)$ is the hypothesized difference between the two population means. When the null hypothesis is that the two population means are equal (the most common case), $(\mu_a - \mu_b)$ is zero. The term s_p is a standard error for the difference between means and is based on the pooled estimate of variance of the two samples, and is calculated as shown in the denominator of equation 10.2.

$$t = \frac{(\bar{x}_a - \bar{x}_b) - (\mu_a - \mu_b)}{\sqrt{\dfrac{s^2_a}{n_a} + \dfrac{s^2_b}{n_b}}} \qquad (10.2)$$

The expression in the denominator of equation 10.2 is the standard error of the difference between the two means (s_p). The degrees of freedom for this statistic are $n_a - 1$ or $n_b - 1$, whichever is the smaller of the two.[1]

Assumptions of the Test

1. Samples are collected randomly from the two populations of interest.
2. The variable is measured on a continuous scale; or, if the measured variable is discrete, then it must assume a large range of possible values.
3. The variable is approximately normally distributed.

The t test for independent samples is both powerful and robust. Recall that statistical **power** means that the probability of a type II error is low; in other words, when the null hypothesis is false, we have a good chance that we will reject it (section 6.3). A test is **robust** if it is still valid when the characteristics of the data depart somewhat from the assumptions. As long as the variable consists of measurements (i.e., not ranks or categories), the t test is generally valid when sample sizes are large. However, the power of the test becomes poor when the failure of the assumptions is severe. When assumptions 2 or 3 are not met, the Mann-Whitney U test (section 10.3) might be more appropriate. To conduct the two sample t test, the test statistic t is computed by equation 10.2.

Example 10.1
A *t* Test: Does Root Hair Morphology Differ between Plant Species?

The ratio of length to width of root hair cells in two species of plants (A and B) of the same genus were measured using random samples from each species. Assume this variable is approximately normally distributed. We wish to know if the two population means differ. The results were as follows.

	Species A	Species B
n	12	18
$\bar{x}$	1.28	4.43
s^2	0.112	7.072

Data from T. Ruhlman

The null hypothesis for this example is

$$H_0 : \mu_A = \mu_B$$

and the alternative hypothesis is

$$H_1 : \mu_A \neq \mu_B$$

Because of the way the question is stated (do the two population means differ?), this is a two-tailed test. We will reject the null hypothesis if the absolute value of our calculated t statistic exceeds the critical value. (Recall the rejection region for an example 2-tailed test in fig. 9.5.) Degrees of freedom here are based on the smaller sample size, so df = 12 − 1 = 11. The critical value of t for 11 degrees of freedom and $\alpha = 0.05$ is 2.201 (table A.2). Next, we calculate the t statistic. Substituting the values from example 10.1 into equation 10.2 gives

$$t = \frac{1.28 - 4.43}{\sqrt{\dfrac{0.112}{12} + \dfrac{7.072}{18}}} = -4.967$$

The absolute value of our calculated t statistic is 4.967. Comparing this value against table A.2 at 11 degrees of freedom indicates the p-value is 0.0001 < p < 0.001. In other words, if the null hypothesis were true, then the chance of obtaining a statistic this large or larger is less than one in a thousand. We therefore reject the H_0 that the two population means are equal and conclude that the mean root hair length ratios for the two species of plant are not equal.

Statistical software packages may sometimes present more than one version of the t statistic. Another version of the t statistic can be calculated when we assume the two population variances are equal. Since the procedures for testing equality of variance for small samples are not very powerful, many researchers prefer to use the method described above. The formulas we present for computing the pooled estimate of the variance and degrees of freedom provide the most conservative approach for conducting the t test. In other words, this approach makes it more difficult to reject the null hypothesis. However, the methods used here and by statistical software will most often reach the same conclusions.

10.2 Confidence Interval for the Difference between Two Population Means

Earlier (section 9.3) we calculated and interpreted a confidence interval for a single population mean. The same approach is also useful for determining the interval which likely brackets the difference between two means. The 95% confidence interval for $\mu_a - \mu_b$ is:

$$(\bar{x}_a - \bar{x}_b) \pm (t_{.05,k}) * \sqrt{\frac{s_a^2}{n_a} + \frac{s_b^2}{n_b}} \qquad (10.3)$$

where k is the degrees of freedom, the smaller of $n_a - 1$ or $n_b - 1$. Substituting the root hair statistics in equation 10.3 (and reversing the order of the groups to give a positive difference)[2] gives:

$$95\% \text{ CI for } \mu_a - \mu_b = (4.43 - 1.28) \pm 2.201 * \sqrt{\frac{7.072}{18} + \frac{0.112}{12}}$$
$$= 3.15 \pm 2.201 * \sqrt{0.4022}$$
$$= 3.15 \pm 1.40 = 1.75, \; 4.55$$

We are 95% sure that the mean difference between groups B and A in their root hair cell length-to-width ratios is at least 1.75 and as great as 4.55. Not only have we demonstrated a difference between groups in their root hair characteristics (earlier hypothesis test), but we also know the magnitude of this difference, with a specified certainty (95%).

Computer statistical packages can easily be used to do two-sample t tests and compute their confidence interval. An example from *Minitab* follows, using the mosquito fish data (Digital Appendix 2).

Interpret this example on your own. Which group appears to be larger? (Inspect the sample means.) By how much? (Inspect the confidence interval and say that the difference is "at least ___ and as much as ___.") What is meant by a 95% confidence interval? Be careful with your answer (refer to sections 6.1 and 9.3).

10.3 A Nonparametric Test for Two Independent Samples: The Mann-Whitney U Test

There are a number of situations in which the data collected in a survey or experiment do not meet the assumptions of the t test for independent samples. For instance, the data might consist solely of ranks (nor measurements) or the distribution of a measurement variable might not be normal. In such cases one is well advised to use the Mann-Whitney U test, which is the nonparametric counterpart of the t test for two independent samples. When the assumptions of the t test are not met, the Mann-Whitney U test is more powerful. On the other hand, when the assumptions of the t test are met, the t test is more powerful (hence the reason why we don't just always use the Mann-Whitney U test!).

The Mann-Whitney U test examines whether two samples could have been drawn from identical populations. Specifically, it tests whether two populations of the same but unspecified distribution differ with respect to central tendency. We'll typically use the median (θ, "theta") for our measure of interest, since it works well with skewed distributions (section 4.2). The null hypothesis of the Mann-Whitney U test is that the samples were drawn from populations with an identical median. In other words:

$$H_0: \theta_1 = \theta_2$$

The assumptions of the test are that the two population distributions are of the same shape (but not necessarily normal) and that random samples have been drawn from the two populations. The sample sizes need not be equal.

To conduct the Mann-Whitney U test, the data of the two samples are ranked together, while at the same time identity of the sample to which each datum belongs is preserved. The lowest value in either sample receives a rank of 1, the next lowest the rank of 2, and so on. Tied scores receive the average

Table 10.1 Comparison of Male and Female Mosquito Fish in their Mean Length (data from Digital Appendix 2). 1 = male, 2 = female. (Output from *Minitab* ver. 16.)

Two-Sample *t* Test and Confidence Interval

```
Two sample T for Mosquitofish_Length

Gender      N       Mean      StDev    SE Mean
1         854 ·     23.60      2.64      0.090
2         797       34.29      5.50      0.19

95% CI for mu (1) - mu (2): ( -11.115,  -10.27)
T-Test mu (1) = mu (2) (vs not =): T= -49.83  P=0.0000  DF= 1126
```

rank that each would have, had they not been tied. It is not as complicated as it sounds. Example 10.2 will help clarify this. To aid in ranking, we first sorted the data in each group.

Example 10.2
The Mann-Whitney U Test

Male gully cats are territorial; they hold territories up to several hectares in size. The territory size of random samples of gully cats from two locations were measured (in hectares) with the results shown as follows. We wish to know if there is a difference in territory size between these two populations. A glance at the raw numbers suggests that the average (median) territory size at location B is greater than that at location A. But we must run a formal hypothesis test to support our intuition.

Location A		Location B	
Territory size	Rank	Territory size	Rank
7	1.5	8	3
7	1.5	10	4.5
10	4.5	18	8
14	6	21	11
17	7	29	14
20.6	9	32	15
21	11	35	16
21	11	36	17
24	13	37	18
		45	19
$n_a = 9$	$\Sigma R_a = 64.5$	$n_b = 10$	$\Sigma R_b = 125.5$

Why might we choose the Mann-Whitney U test for these data rather than a t test for independent samples? We use this nonparametric test because we suspect that territory size in these imaginary animals is not normally distributed. Furthermore, because the sample sizes are small, the central limit theorem does not help us out.

First, all of the observations in both samples are ranked, with tied observations receiving the average rank that they would have if they were not tied. Note in the "location A" sample that there are two observations of 7, and these are the lowest of all of the observations. If they are not tied, they would receive the ranks 1 and 2. However, since they are tied, they receive the average rank that they would have, or $(1 + 2)/2 = 1.5$. Ranking proceeds in this manner until all observations have been ranked.

We now compute two values of the test statistic, U, by the following equations:

$$U_a = n_a n_b + \frac{n_a(n_a + 1)}{2} - \sum R_a \qquad (10.4)$$

where n_a and n_b are the sample sizes of samples A and B, and ΣR_a is the sum of the ranks of sample A. U_b is then calculated by

$$U_b = n_a n_b - U_a \qquad (10.5)$$

For the example

$$U_a = (9 \times 10) + \frac{9(9+1)}{2} - 64.5 = 70.5$$

and

$$U_b = (9 \times 10) - 70.5 = 19.5$$

Table A.4 gives the critical values of U when both sample sizes are smaller than 20. The table is divided into two parts. For two-sided tests (at $\alpha = 0.05$), use the top half; for one-sided tests (at $\alpha = 0.05$), use the bottom half. The null hypothesis is rejected if either U_a or U_b is equal to or larger than the critical value. In this example, U_a (70.5) is the larger of the calculated values. Checking table A.4 for a two-sided test at $\alpha = 0.05$, and sample sizes of 9 and 10, the critical value of U found in table A.4 is 70. Since our larger calculated U is larger than this, we reject H_0 for the gully cat example, and conclude that territory sizes in the two locations are not equal.

Because of space limitations, table A.4 has critical values for a limited range of alpha. We thus can not determine p-values very precisely. However, the fact that the test statistic (70.5) was just slightly larger than the critical value (70), suggests that the p-value is just less than 0.05. Statistical packages, such as *Minitab*, report an exact p-value. Table 10.2 shows the results from *Minitab* for this same dataset. The test statistic that is reported (W) is simply the sum of the ranks for the first group. Notice that the p-value (0.0407) is right where we expect, less than 0.05, but not by much.

To conduct a *one-sided test*, you would do two things differently from the two-sided test. First, at $\alpha = 0.05$, use the bottom half of table A.4. Second, use either U_a or U_b, depending on the following criteria, which follow from the structure of your hypotheses:

If H_0: $\theta_a \geq \theta_b$ (vs. H_a: $\theta_a < \theta_b$), then use U_a (from eq. 10.4) for the test statistic.

If H_0: $\theta_a \leq \theta_b$ (vs. H_a: $\theta_a > \theta_b$), then use U_b (from eq. 10.5) for the test statistic.

Whenever conducting a one-sided test, it is always a good idea to inspect the data to make sure the result makes sense. If you reject the null hypothesis and conclude that one group has a larger median than a second group, then that first group should tend to have larger values than the second group. Plot your data and do descriptive statistics to visualize this.

Table 10.2   Mann-Whitney U Test with *Minitab*. Data from example 10.2.

Mann-Whitney U Test and CI: LocationA, LocationB

```
                N       Median
LocationA       9        17.00
LocationB      10        30.50

Point estimate for ETA1-ETA2 is -12.50
95.5 Percent CI for ETA1-ETA2 is (-22.01,0.01)
W = 64.5
Test of ETA1 = ETA2 vs ETA1 not = ETA2 is significant at 0.0412
The test is significant at 0.0407 (adjusted for ties)
```

Note that table A.4 applies only when sample sizes are 20 or less. If each sample has more than 20 observations, the probability distribution of U is approximately normal. This allows us to calculate z and use the table for the normal distribution (table A.1). To calculate z, use equation 9.6 below.

$$z = \frac{\left[U - \left(\frac{n_a n_b}{2}\right)\right]}{\sqrt{\frac{n_a n_b (n_a + n_b + 1)}{12}}} \qquad (10.6)$$

U is either U_a or U_b, calculated according to equations 10.4 and 10.5. The denominator in equation 10.6 is the standard error of U. H_0 is rejected at $\alpha = 0.05$ if $|z| \geq 1.96$ for a two-sided test.

10.4 Power of the Test: How Large a Sample Is Sufficient?

Consider the following situation. An industry releasing pollutant X into a stream would like to show government regulators that they are having no effect on concentrations of X in the stream. They already know that background levels of X are naturally quite variable. Their employees collect five water samples in stream locations upstream and five downstream from the discharge pipe, and measure concentrations of X in each sample. Although the calculated sample mean downstream from the pipe is considerably greater than the mean upstream from the pipe, the variance for each group is large. The calculated t statistic is small and thus H_0 (no difference) cannot be rejected. The industry reports to the regulator that there is no significant difference and thus they are having no clear effect on the stream. There is something fishy here. What is it? Of course—the power of their test was small, making it exceedingly difficult to reject H_0 even when it's actually false!

In previous sections we have brought up the term "power." The **power of the test** is the chance of rejecting the null hypothesis when it is indeed false. Recall that power is equivalent to $1 - \beta$ (one minus the type II error rate). Although power is not yet commonly analyzed in biological statistics, researchers are now encouraged to make wider use of power analysis.

In general, the power of a test is influenced by three properties that we can control:

1. The type I error rate (α);
2. The difference between two means which we can discriminate ("**minimum detectable difference**," also called "**effect size**") (δ); and
3. Sample size (n)

In most cases, we set α to some fixed value, say 0.05. If we were to increase α to a higher error rate (say, 0.10), then β would decline and power would go up. Increasing α is not always desirable. If the difference between the means is quite large, then we can more easily discriminate them than if the difference is small. In other words, our t test would more likely reveal a significant difference when there is indeed a large difference between means. Finally, all things being equal, as sample size (n) goes up, power increases. This is because the sampling distributions for each of the means becomes narrower with increased sample size (section 8.3). The net effect is that, with larger samples, we are more likely to discriminate a real difference between groups. A simple graphic illustrates this effect (fig. 10.1).

More formally, power analysis allows us to determine one of several properties of a sampling distribution. For instance, for a given level of α and sample size (n), we can determine power ($1 - \beta$) for a range of differences between the means of two groups. More commonly, we set power to some desired level and determine one of the two remaining properties.

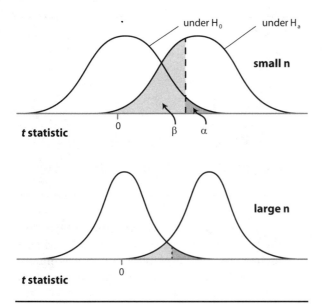

Figure 10.1 An illustration of how the power of the test improves with increased sample size. The distribution of the t statistic is shown, both when H_0 is true and when it is false. When H_0 is true, a value of t beyond a particular decision criterion (dashed vertical line) leads to a type I error with probability alpha (hatched area). When H_0 is false, the distribution of the t statistic for a particular alternative hypothesis is also shown. Values of t to the left of the decision criterion lead to accepting H_0 (a type II error), with probability beta (dotted area). Power of the test is 1-beta, the area under H_a to the right of the decision criterion. When sample size is large, the distribution of the t statistic is narrower (greater precision) and the areas of overlap between the two curves is less. Since we keep alpha constant (usually 0.05), beta is smaller and hence power is greater.

10.4.1 Determining the Sample Size Needed to Detect a Minimum Effect

First, if we have some preliminary data, we can use power analysis to determine sample size needed to detect a minimum difference between two groups. For example, we might want to be able to detect a blood pressure change of at least 5 mm Hg following a drug treatment. Preliminary data are needed to provide an estimate on the variance. By estimating the required sample sizes, we then have some guidance for designing an effective future experiment.

If we know the desired power and specify a minimum difference (effect size) between two groups (δ), we can determine the sample size needed by equation 10.7. In order to do this power analysis, we must have a prior estimate of the vari-

ance and assume that both populations are normally distributed. When both populations have the same variance, estimated by s_p^2, and sample sizes (n) are equal for each group, and the degrees of freedom (γ) = $2(n-1)$, the required sample size for a 2-sided t test is:

$$n \approx \frac{2s_p^2}{\delta^2} * (t_{\alpha,\gamma} + t_{2\beta,\gamma})^2 \qquad (10.7)$$

An example should help illustrate.

Example 10.3
A Power Analysis

Milk production (in kg/day) of dairy cattle was compared between two groups, those getting a vitamin supplement and those not getting the supplement. In a preliminary study, milk production was shown to be normally distributed with a variance of 0.64. For a 2-sided test with $\alpha = 0.05$, what sample size is necessary to detect a difference of at least 0.5 kg/day, while keeping the power at least 90%?

Power = $1 - \beta \geq 0.9$, so $\beta \leq 0.10$. We thus use $\beta = 0.10$ in our formula. Since the critical values for t depends on degrees of freedom, which we don't know, we'll first choose the critical values from the last row of the table (∞). $t_{0.05,\infty} = 1.960$ and $t_{0.20,\infty} = 1.282$. Now, substituting into the general formula:

$$n \approx \frac{2(0.64)}{(0.5)^2} * (1.96 + 1.282)^2 = 53.8 \approx 54$$

In other words, at least 54 cows should be included in each group. If we refer back to the t table, the critical values for $2(54 - 1) = 106$ degrees of freedom are close to the original numbers we picked, so the calculation should come pretty close to what we originally got. $t_{0.05,106} = 1.99$ and $t_{0.20,106} = 1.29$. Now, substituting again:

$$n \approx \frac{2(0.64)}{(0.5)^2} * (1.99 + 1.29)^2 = 55.1$$

To be safe, we should use at least 56 cows in each group in our next experiment.

10.4.2 Determining the Minimum Detectable Difference

Suppose we are interested in how large a difference we could detect in a study which is already completed. If we know the sample size and variance and specify a particular power level, we can determine the minimum detectable difference (effect size, δ) by equation 10.8.

$$\delta \geq \sqrt{\frac{2(s_p^2)}{n}} * \left(t_{\alpha,\gamma} + t_{2\beta,\gamma}\right) \qquad (10.8)$$

Continuing with our example from the dairy cattle, imagine that we had obtained samples of size 30 from each group and we wanted to keep power of at least 90%. Substituting into this equation the appropriate critical values ($t_{.05,\,29} = 2.045$, $t_{.20,\,29} = 1.311$) and statistics ($s_p^2 = 0.64$, $n = 30$), we get:

$$\delta \geq \sqrt{\frac{2(0.64)}{30}} * (2.045 + 1.311) =$$

$$\delta \geq (0.2066) * (2.045 + 1.311) = 0.69 \text{ kg/day}$$

In other words, in this study of dairy cattle, the difference in milk yields between the vitamin-supplement and control groups would have to be at least 0.69 kg in order for us to detect a significant treatment effect.

In conclusion, power analysis can be a useful tool for experimental design, by allowing us to determine necessary sample sizes for a future experiment. Power analysis also provides a method for showing how small an effect could have been detected by a particular experiment. This is particularly important when we find no significant difference between treatment groups. The power analysis allows others to see how sensitive our experiment was. In the case of the stream pollution described earlier, we would likely have found that the minimum detectable difference was quite large. A regulatory agency would do well to require a more sensitive study.

Some statistical reference books (e.g. Sokal and Rohlf 1995, Zar 1999) provide detailed graphics to aid in power determinations for various statistical tests. Some computer packages, such as *Minitab 16*, also do power analysis.

10.5 Review: What Is the Appropriate Statistical Test?

In the last two chapters, you have been introduced to a variety of tests for making inferences about means (or medians). Let's review the rationale for choosing the appropriate statistical test, which depends on answers to the following questions:

1. How many populations are we considering?
2. If more than one population, are samples related or independent?
3. Are the data continuous measurements?
4. Are the data normally distributed? If not, are sample sizes large ($n > 30$)?

Recall that parametric statistics (like the *t* test) require continuous measurements and either a normal distribution or large sample sizes. For comparisons between two treatments, the following graphic may serve as a helpful key for choosing the correct test (fig. 10.2).

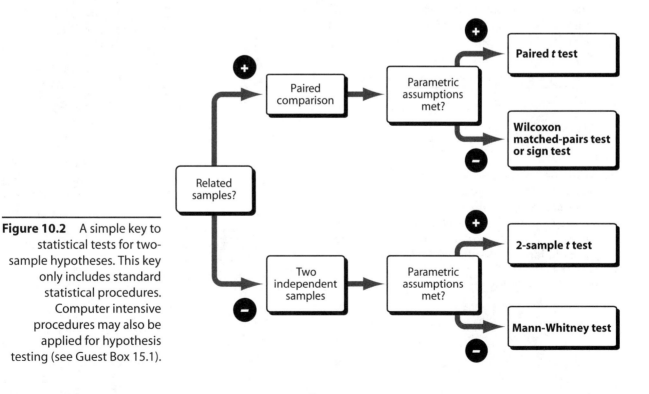

Figure 10.2 A simple key to statistical tests for two-sample hypotheses. This key only includes standard statistical procedures. Computer intensive procedures may also be applied for hypothesis testing (see Guest Box 15.1).

Key Terms

F-statistic
error variance
independent samples
Mann-Whitney U test
minimum detectable difference (= effect size)
power of the test
related samples
robust
t test

Exercises

For each of the following problems, state the appropriate null hypothesis and the alternative hypothesis, conduct the appropriate test, and state a conclusion. Reject the H_0 if $p \leq 0.05$. Be sure to include the p-value. Assume the variable in question to be approximately normally distributed unless instructed otherwise.

THE *t* TEST FOR INDEPENDENT SAMPLES

10.1 Random samples of cranberries were collected from two bogs. We wish to know if the mean weight of cranberries (in grams) differs between the two bogs. The results were as follows.

	Bog A	Bog B
$\bar{x}$	1.31	1.27
s	0.25	0.27
n	25	27

10.2 Random samples of largemouth bass and smallmouth bass were taken from a lake, and their lengths (in millimeters) were determined. We wish to know if the mean standard length differs between the two species in this lake. The results were as follows:

	Largemouth Bass	Smallmouth Bass
$\bar{x}$	272.8	164.8
s	96.4	40.0
n	125	97

Data from J. Kagel

10.3 Data on resting pulse rates (in bpm) were collected for random samples of 57 men and 63 women between the ages of 18 and 21. We wish to know if there is a difference in the mean pulse rate of men and women in this population. The results were as follows.

	Men	Women
$\bar{x}$	73.789	82.270
s	10.395	13.750
n	57	63

a. Use a two-sample *t* test to examine this hypothesis.
b. Compute a 95% confidence interval for $\mu_m - \mu_f$.
c. Interpret your results for part a and b. Are the results consistent? Explain.

10.4 Data on reaction time (in milliseconds) for random samples of 58 men and 68 women were collected. We wish to know if there is a difference in reaction time between men and women in this population. The results were as follows.

	Men	Women
$\bar{x}$	170.21	181.31
s	32.643	45.988
n	58	68

a. Use a two-sample *t* test to examine this hypothesis.
b. Compute a 95% confidence interval for $\mu_m - \mu_f$.
c. Interpret your results for part a and b. Are the results consistent? Explain.

10.5 Liver alcohol dehydrogenase activity in two random samples of catfish was determined. One sample was taken upstream from a brewery, and the other sample was taken downstream from the brewery. We wish to know if living downstream from a brewery increases liver alcohol dehydrogenase in these animals. The results were as follows.

Upstream	Downstream
10	30
25	32
8	28
11	35
19	29
7	32
5	32
30	38
	31

10.6 The effect of copper sulfate on the mucus cells in the gill filaments of a certain species of fish was investigated. We wish to know if exposure for 24 hours to copper sulfate reduces the number of mucus cells in this species. The number of mucus cells per square mm in the gill filaments of both untreated fish and exposed fish were as follows.

Untreated	Exposed
16	10
17	8
12	10
18	12
11	13
18	14
12	6
15	5
16	7
14	5
18	10
12	11
	9
	8

10.7 A certain species of bacterium was grown with either glucose or sucrose as a carbon source. After a period of incubation, the number of cells ($\times 10^6$) was determined. Is there a difference in growth rate of the bacterium between the two carbon sources?

Glucose	Sucrose
6.3	5.8
5.7	6.2
6.8	6.0
6.1	5.1
5.2	5.8

10.8 Six randomly selected pea plants were treated with a plant growth regulator, and six randomly selected plants were not treated. We wish to know if the growth regulator affects internode growth.

Internode Length (in millimeters)

Treated Plants	Untreated Plants
15.2	13.5
12.3	9.8
11.6	10.2
14.8	8.7
10.0	9.2
14.2	9.0

10.9 Using the data in Digital Appendix 3, determine if male athletes have faster reaction times than male nonathletes.

10.10 Using the data in table Digital Appendix 3, determine if women smokers have faster pulse rates than women nonsmokers.

THE MANN-WHITNEY U TEST

10.11 Do people find hairy spiders scarier than nonhairy spiders? To find out, 20 people were randomly assigned to two groups of 10 each. One group viewed a hairy spider, and the other group viewed a very similar but non-hairy spider. Each person was asked to rate the spider she or he viewed on a scariness scale from 1 to 10 (10 being most scary). The results were as follows.

Hairy		Nonhairy	
10	10	7	5
8	9	6	4
7	9	8	5
9	5	6	6
9	8	1	3

10.12 Male hoop snakes, upon encountering one another, may engage in a protracted ritualized combat behavior until one establishes himself as dominant over the other. We would like to know if these encounters last longer in the presence of a female. 24 males were randomly assigned to pairs. 6 randomly selected pairs were tested in the presence of a female, and 6 were tested in the absence of a female. This variable is probably not normally distributed. The results were as follows.

Interaction Time (in minutes)

Pairs without Female	Pairs with Female
10	59
15	35
8	70
30	65
1	43
80	90

10.13 The 72-hour blastogenesis of chicken peripheral blood lymphocytes from a group treated with PHA and from an untreated group are given below. This variable is *not* approximately normally distributed. We wish to know if treatment with PHA has an effect on blastogenesis of these cells.

Control	Treated
1631	87700
50102	69553
1369	76215
41188	366
387	40104
498	38661
259	141153
329	154805
4330	123075
5002	627
658	126175
300	11223
	300

10.14 Seven tomato plants were treated with chlorogenic acid to determine if this would influence the activity of the enzyme o-diphenol oxidase in the leaves. A control group of seven plants were not treated. We do not know if this variable is approximately normally distributed, nor is it possible to determine this with a sample as small as this. Does this treatment affect activity of the enzyme?

Treated	Untreated
35	10
45	18
36	8
11	29
41	17
29	8
38	11

TEST YOURSELF

Choose and conduct the appropriate statistical test covered in chapter 9 or chapter 10. Answer all these exercises.

10.15 Aerial surveys of several randomly selected areas of forest land were used to determine damage by a certain insect. Some areas had been sprayed several years before the survey to control the insect and some had never been sprayed. A rank scale of 1 to 10 was used to assess damage, with 10 being most severe. We wish to know if there is a difference in previously sprayed areas and in areas that had never been sprayed.

Sprayed Areas	Unsprayed Areas
3	2
0	5
1	6
5	3
2	3
1	4
5	8
3	2
6	1
0	8
2	
6	
5	

10.16 *Brucella abortus* antibody titers (pfc/10^6 cells) in 15 turkeys were measured before and after a period of stress. We wish to know if stress affects antibody titer.

Turkey Number	Before Stress	After Stress
1	20	17
2	18	14
3	19	16
4	18	19
5	17	14
6	14	18
7	17	8
8	10	10
9	13	12
10	16	15
11	20	8
12	17	6
13	16	17
14	19	5
15	8	3

10.17 Spleen weights were measured from road warblers infected (i) with the avian malaria parasite and of those that were parasite free (h). We wish to know if infection by this parasite affects spleen weight in this species. Also compute a 95% confidence interval for $\mu_i - \mu_h$.

Infected	Healthy
25.6	20.8
27.8	22.9
29.3	26.0
26.9	23.2
26.0	25.1
25.9	23.7
	25.6
	23.2

10.18 High concentrations of phosphate can contribute to excessive growth of algae, which can be harmful to freshwater fish. Aquarium suppliers recommend that the phosphate level does not exceed 0.5 ppm in a freshwater aquarium. The data below show the results of ten random samples taken from a large freshwater aquarium. Is the phosphate level safe?

0.45	0.49
0.58	0.54
0.51	0.52
0.52	0.52
0.49	0.48

Notes

[1] Other methods of computing degrees of freedom may produce higher values, depending on equality of variances and sample sizes between groups. We have chosen a conservative approach, which is also simpler and more direct to calculate. Those using *Minitab* will see higher values of degrees of freedom, based on the following equation:

$$df = \frac{\left(\dfrac{s_a^2}{n_a} + \dfrac{s_b^2}{n_b}\right)^2}{\left(\dfrac{\left(\dfrac{s_a^2}{n_a}\right)^2}{n_a - 1} + \dfrac{\left(\dfrac{s_b^2}{n_b}\right)^2}{n_b - 1}\right)}$$

[2] Since we had a 2-sided hypothesis, such a reversal is quite acceptable for improving clarity. However, when evaluating 1-sided hypotheses, be very careful with the sequence of groups; the sequence in the equation should be the same as in the hypothesis.

chapter eleven

Inferences Concerning Means from Multiple Populations
ANOVA

Analysis of variance (ANOVA) is one of the most versatile and useful techniques of statistical inference. ANOVA is the statistical tool for analyzing the effects of different groups (a categorical variable) on a particular measurement variable. Generally, we use ANOVA to answer the question: "Are the means from several groups all the same?" In this sense, ANOVA is an extension of the two-sample problem (chapter 10), although it uses a different approach. Essentially, ANOVA is a technique of partitioning the variance in a set of data into several components in such a way that the contribution of each of these components to the overall dataset may be assessed. The techniques of ANOVA are useful both in situations where the researcher can carefully design and control experiments, and also for the analysis of certain types of observational (survey) data. One of the strengths of ANOVA is that it can be extended to a wide variety of experimental designs. In the next two chapters, we will examine the basic concepts and most common applications of ANOVA. In the present chapter we will explore the use of ANOVA to detect effects of groups of a single factor on the measurement (**one-way ANOVA**). We will illustrate calculation steps for situations where sample sizes are equal within each group. With slight modifications, ANOVA can also be used with unequal sample sizes ("unbalanced data"), although space does not allow us to show examples of these cases. Since most investigators use computer software to do

ANOVA, we include *Minitab* output to compare with our worked examples.

11.1 The Rationale of ANOVA: An Illustration

The rationale of analyzing the variance of samples when our interest is in their means may be illustrated by this hypothetical example. Fifteen juniper pythons, all of similar age and size, were randomly assigned to one of three groups. Group A received one drug; group B received another drug; and group C received a placebo. The heart rate of each snake was then determined to see if either or both of the drugs affects heart rate. The null hypothesis in an experiment such as this is that the means of all three populations are equal, or

$$H_0: \mu_A = \mu_B = \mu_C$$

Population, in this case, refers to the heart rate response of all possible juniper pythons that might be given a particular treatment.

One approach for solving a case like this would be to conduct a series of two-sample t tests, comparing each group with each other group (i.e., group A with group B, A with C, and B with C). There are two general problems with using this approach. The first reason is it's not very efficient. To demonstrate, imagine that we instead had 10 different treatments we were comparing. Making all possible comparisons of pairs among 10 treatments would require 45 different *t* tests! Running 45 tests would be quite time consuming. A second problem is that so many tests increases the type I error rate. Suppose we set alpha at 0.05 for rejecting H_0 in any individual *t* test (i.e., we will reject the null hypothesis that the means of any two groups are equal when $p \leq 0.05$). In such a situation there is a probability of 0.05 of making a type I error when comparing any two sample means. If we compared all possible group means for 10 groups (45 *t* tests), the chance of a type I error would be far greater (about 0.90), causing us to incorrectly reject the null hypothesis that all of the group means are equal. Analysis of variance is designed to overcome this problem.

Let's return to our imaginary example with the three groups of juniper pythons. Because all of the individual juniper pythons in the study were chosen from the same population or group, we may assume that all of the individuals (the "experimental units") involved in the experiment are more or less homogeneous. Any variation from one individual to another within a group is the normal variation we expect among individuals in a population. This variation is called **within-groups variance** (also called **error variance**). Because snakes from the available group of 15 were randomly assigned to the three treatment groups, we expect the error variance within each group to be approximately the same. Suppose we obtained the results shown in table 11.1.

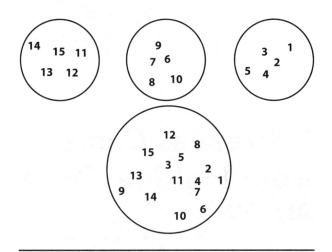

Figure 11.1 Thought experiment using the juniper python heart rate data. The within-groups variance is small for each group (top). Imagine if all the measurements were mixed in one big group (bottom); here the variance would be much larger.

this variance (s_t^2) is considerably larger. (Try making the calculations to verify.)

The approach with ANOVA is to partition the variance in a set of data into various components and to determine the contribution of each of these components to the overall variation. First, we consider the overall variation in the complete dataset (all 15 snakes considered together). This is called the total variance, and it is about 20.00 in this case. There are two sources of this variation. One source is the usual variation from one individual to another, caused by genetic and environmental differences among individuals, and it has nothing to do with the treatments the animals were given. This is called the within-groups variance (error variance). Note in table 11.1 that the error variance in each group is the same (about 2.5). There is quite a difference between this error variance and the total variance, which tells us there is a source of variance that is not accounted for by the variation from one individual to another. Its source is the variance introduced by our treatments. The difference between the total variance in the dataset and the within-groups variance is called the **among-groups variance** (also called **treatment variance**).

The sampling distribution that describes the ratio of these two variances, or

$$\frac{\text{among-groups variance}}{\text{within-groups variance}}$$

is the **F distribution**. The *F* distribution is another probability distribution, the shape of which depends on two types of degrees of freedom (df), those asso-

Table 11.1 The Effect of Two Drugs (A and B) and a Placebo Control (C) on the Heart Rate of Juniper Pythons (in beats per minute). This example demonstrates a difference among groups.

Group A	Group B	Group C
13	7	3
15	6	5
12	10	2
14	9	4
11	8	1
$\bar{x} = 13$	$\bar{x} = 8$	$\bar{x} = 3$
$s^2 = 2.50$	$s^2 = 2.50$	$s^2 = 2.50$
	$s_t^2 = 20.00$	

Here the error variance is the same for each group (2.50). Now imagine that all 15 measurements were lumped together (fig. 11.1). Because some measurements are quite different from others,

ciated with the numerator and those associated with the denominator of the above expression. The among-groups degrees of freedom is determined as the number of treatments minus one and the within-groups degrees of freedom is the total number of observations in all groups minus the number of groups. A typical F distribution is shown in figure 11.2. The shaded portion of the curve is 0.05 of the total, and it is termed alpha (α), as before. The values of F that delimit this alpha region are found in table A.6, whose use will be described later.

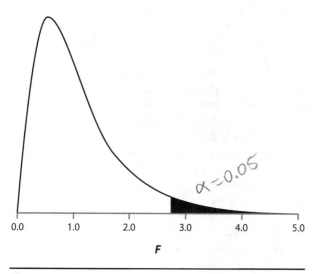

Figure 11.2 An F distribution. The shaded area illustrates alpha.

When the treatments have no effect, virtually all of the variance in the dataset will be due to within-groups variance which, you will recall, is the normal variance among individuals in a population. In this case, the variance ratio will be low (generally close to 1.0). When treatments have a real effect, the variance ratio will be much larger (i.e., $F \gg 1.0$).

Consider another outcome of this same imaginary experiment, shown in table 11.2. In this case the error variance and the total variance are practically the same, and we would probably conclude that most of the variation within the dataset is attributable to the usual variation between individuals and that little or none of it results from the treatments given. In this case we would no doubt fail to reject H_0 and conclude that neither drug has an effect on heart rate. How we reach this decision is explained in the following sections.

Table 11.2 The Effect of Two Drugs (A and B) and a Control (C) on the Heart Rate of Juniper Pythons (in Beats per Minute). This example demonstrates no difference among groups.

	Group A	Group B	Group C
	1	3	2
	3	5	1
	2	1	5
	5	4	4
	4	2	3
$\bar{x}$	3	3	3
s^2	2.50	2.50	2.50
		$s_t^2 = 2.14$	

11.2 The Assumptions of ANOVA

There are several important assumptions of ANOVA, which should be adhered to rather closely. When these assumptions are not met, we cannot trust the results of the analysis. (Theorists tell us that alpha is not what we believe it to be.)

Assumptions of ANOVA
1. Each of the groups is a random sample from the population of interest.
2. The measured variable is continuous (or if discrete, it may assume a large range of values).
3. The error variances are equal.
4. The variable is approximately normally distributed.

Assumption 1 may not be violated! Randomization is usually achieved by randomly assigning the available experimental subjects to the various treatment groups, as in the juniper python illustration. Failure of assumption 1 cannot be fixed; we might as well throw out the data. When any of assumptions 2 through 4 are not met, one should consider use of a nonparametric method, discussed later in this chapter (section 11.5.2). The assumption that the error variances are equal (homogeneous) should always be checked. Later in this chapter (section 11.4) we will see how this is done.

A variety of experimental designs can be analyzed with ANOVA, and each of these requires a somewhat different set of computations. Which design is chosen depends on the nature of the question being asked. In the following sections, we will examine different forms of one-way ANOVA and its nonparametric analogue, the Kruskal-Wallis test. Other designs are introduced in chapter 12.

11.3 Fixed-Effects ANOVA (Model 1)

Example 11.1 is an imaginary but fairly typical situation in which the fixed-effects ANOVA is useful. There are three sample means representing three treatments. In this case, we wished to know if either or both of the two drug treatments are significantly different from the control. These are random samples because the individual subjects (experimental units) were randomly assigned to the three groups.

Example 11.1
An Experiment Using the Fixed-Effects Design

A group of 30 highly inbred mice, just weaned, were randomly divided into 3 groups of 10 mice each and were given three different diets. At the end of several weeks, the gain in weight of each mouse was determined. Group 1 was fed regular laboratory mouse food (the control diet); group 2 was fed potato chips, Twinkies, and diet cola (the junk food diet); and group 3 was fed granola and organically grown prune juice (the health food diet). The results of the experiment are shown in table 11.3. We wish to determine if the mean weight gain in any of the three treatments (groups) is significantly different from the other means. Suppose the results of our imaginary experiment were as follows.

Weight Gain in Mice

Group 1 (Control)	Group 2 (Junk Food)	Group 3 (Health Food)
10.8	12.7	9.8
11.0	13.9	8.6
9.7	11.8	8.0
10.1	13.0	7.5
11.2	11.0	9.0
9.8	10.9	10.0
10.5	13.6	8.1
9.5	10.9	7.8
10.0	11.5	7.9
10.2	12.8	9.1
$\bar{x}_1 = 10.28$	$\bar{x}_2 = 12.21$	$\bar{x}_3 = 8.58$
$n_1 = 10$	$n_2 = 10$	$n_3 = 10$

Inspection of the sample means above suggests that the population means are not all the same. However, a statistical analysis using ANOVA is needed to verify that our intuition is correct. The null hypothesis for this analysis is:

$$H_0: \mu_1 = \mu_2 = \mu_3$$
(weight gain does not depend on diet)

and the alternative hypothesis is:

$$H_A: \text{not all } \mu\text{'s are equal}$$
(weight gain depends on diet)

An experiment of this type is referred to as a **fixed-effects ANOVA (model I)**. Fixed effects means that the treatments to be used in the experiment are chosen by the investigator. This design may also be called a one-way ANOVA with fixed effects, since each observation (individual) in the dataset is classified according to only one criterion—the group to which it is assigned (in this case, diet). The experimental units (individual mice) are randomly assigned to the various treatment groups by the investigator. Since each experimental unit is from the same population, it is assumed that the variation among individuals within any treatment group is the variation usually expected among individuals in a population.

Before beginning a discussion of how one proceeds to test the null hypothesis, we need to consider how and why this experiment was designed as it was. First, it is helpful in such an experiment if the error variance—the variance not associated with the treatments—can be kept to a minimum. This can often be accomplished by selecting experimental units (mice, in this case) that are as nearly alike as possible. For this experiment a group of inbred mice of the same age was used. Randomly assigning the experimental units to the treatment groups is crucial to ensure that variation among individuals that still exists (in spite of our best efforts to minimize it) is not associated with one group more than with any other, and that the variance within each group (the error variance) will be equal among all the groups.

Once the sample has been properly selected and the experiment begins, it is necessary to ensure that all groups are treated identically, as far as possible, with respect to housing, water, temperature, and in every other way except for the factor being tested (diet, in this case). Only then may we feel confident that any difference in weight gain noted among the three groups is caused by diet and not by some other factor.

11.3.1 Testing the Null Hypothesis That All Treatment Means Are Equal

For any number of groups (a), we can use ANOVA to test the null hypothesis that all of the treatment (group) means are equal, or

$$H_0: \mu_1 = \mu_2 = \ldots = \mu_a$$

Recall that we expect the ratio of the among-groups variance to the within-groups variance (the variance ratio, or F) to be approximately 1.0 when the null hypothesis is true, and that the sampling distribution that describes the variance ratio is the F distribution.

Example 11.1 will be further used to illustrate the calculations for obtaining the variance ratio, F, where sample sizes are equal within each group. The data from this experiment are repeated and preliminary calculations needed to compute the variance ratio are given in table 11.3. We will describe these calculations step by step.

Preliminary calculations for example 11.1: These calculations are used to determine the various sums of squares, which will be later used to determine variances. The variances, in turn, are used to determine the variance ratio (F), discussed above. Notice that the steps above have numbers to correspond to the descriptions below.

Tips for the wise: These calculations can be a bit tedious, so some tricks will save you much time and anxiety! First, notice that the layout of the table above resembles a spreadsheet. A computer spreadsheet is the simplest way to do the calculations, since many of the calculations involve doing sums. If you use a calculator, make liberal use of the memories, since the sums can be easily recalled. Finally, as a check on your answers, all variance terms you compute should be positive numbers.

1. Sum the observations for each group and then sum across all groups to get a **grand total**. The grand total may also be coded as $\Sigma\Sigma X$. In the current example, $\Sigma\Sigma X = 310.7$.

2. Sum the squared observations for each group and sum across all groups. In the current example, this total, $\Sigma\Sigma X^2 = 3305.09$. Notice the difference between this set of operations and the operation in step 4 below.

3. Square each of the group totals, divide by n, and then sum across all groups. For instance, in the current example for the first group, 102.8 squared equals 10,567.84, and then divided by the sample size for the group ($n = 10$) equals 1,056.784. Summing across all three groups gives 3283.789.

4. Square the grand total and divide by the total number of observations from all groups. This result is sometimes called the **correction term**. In the current example, where sample sizes within each group are equal, the total number of observations ($a*n$) = $3*10$ = 30. So, 310.7 squared and then divided by 30 is 3217.816333 or about 3217.816.

5. Now we are ready to subtract to obtain the various sums of squares terms. To determine the **total sum of squares**, subtract the correction term from the sum of squared observations. In other words, **step 2 – step 4**, which in this case would be 3305.09 – 3217.816 = 87.274.

Table 11.3 Preliminary Calculations for Weight Gain of Mice fed Different Diets. Group 1—Control, Group 2—Junk Food, and Group 3—Health Food. Data from example 11.1. In this example, the number of groups (a) is 3 and the sample size within each group (n) is 10.

	Weight Gain in Mice				
	Group 1 (Control)	Group 2 (Junk Food)	Group 3 (Health Food)		
	10.8	12.7	9.8		
	11.0	13.9	8.6		
	9.7	11.8	8.0		
	10.1	13.0	7.5		
	11.2	11.0	9.0		
	9.8	10.9	10.0		
	10.5	13.6	8.1		
	9.5	10.9	7.8		
	10.0	11.5	7.9		
	10.2	12.8	9.1		
Group sums				Grand totals	Step
ΣX	102.8	122.1	85.8	310.7	1
ΣX^2	1059.76	1502.41	742.92	3305.09	2
$(\Sigma X)^2/n$	1056.784	1490.841	736.164	3283.789	3
			$(\Sigma\Sigma X)^2/an$	3217.816	4
			SS total	87.274	5
			SS among	65.973	6
			SS within	21.301	7

6. Determine the **sum of squares among groups** by subtracting the correction term from the answer to step 3. In other words, **step 3 – step 4**, which in this case would be 3283.789 – 3217.816 = 65.973.

7. Finally, determine the **sum of squares within groups** by subtracting the sum of squares among groups from the total sum of squares. In other words, **step 5 – step 6**, which in this case would be 87.274 – 65.973 = 21.301.

To summarize, we obtain the sums of squares as follows:

$$SS_{Total} = \sum^{a}\sum^{n} x^2 - \frac{(\sum\sum x)^2}{an} \quad \text{(11.1)}$$

$$SS_{Among} = \sum^{a} \frac{(\sum^{n} x)^2}{n} - \frac{(\sum\sum x)^2}{an} \quad \text{(11.2)}$$

$$SS_{Within} = SS_{Total} - SS_{Among} \quad \text{(11.3)}$$

We next can determine the different variance terms. In ANOVA, the variance is also called the **mean square** (MS). To convert these sums of squares into mean squares, we must divide each by their appropriate degrees of freedom. The among-groups degrees of freedom (df_a) is the number of groups minus one ($a - 1$). The total degrees of freedom (df_t) is the total number of measurements minus one ($an - 1$), and the within-groups degrees of freedom (df_w) is obtained by subtracting the among-groups degrees of freedom from the total degrees of freedom. We usually arrange the sums of squares, degrees of freedom, and mean squares in a tabular form like tables 11.4 and 11.5. The mean squares of interest are obtained by dividing the sums of squares by their appropriate degrees of

freedom. *F* (the variance ratio) is obtained by dividing the among-groups mean square by the within-groups mean square.

Work through the example to ensure comprehension! We now consult table A.6, "Critical Values of the *F* Distribution," to see if we may reject H_0. The top row of the table gives degrees of freedom for the numerator mean square (the among-groups mean square), and the left column gives degrees of freedom for the denominator mean square (the within-groups mean square). In this case the degrees of freedom are 2 and 27. If our calculated value of *F* is greater than or equal to the tabular value of *F*, we will reject H_0. The table value for 2 and 27 degrees of freedom (use 2 and 25, since 2 and 27 is not tabulated) is 3.39. Since our calculated value is much greater than this, we may reject H_0 and conclude that the means of the three treatments are not all equal. Further inspection of table A.6 shows that our calculated *F* statistic also exceeds the table value at $\alpha = 0.01$ (5.57); thus we can express the p-value as $p < 0.01$. In other words, if the null hypothesis were true, then the chance of obtaining a statistic so large would be less than 1%. Such a small value is clearly grounds for rejecting the null hypothesis. We thus conclude that diet does indeed affect weight gain in these mice.

Computer software can run an ANOVA much faster than these hand calculations. If you have software available, run the analysis on this example and confirm that you obtain the same answer in the ANOVA table. An example from *Minitab* appears in table 11.6. Notice in particular the ANOVA table, as well as summary statistics and plots of confidence intervals below this table. We will discuss the Tukey test in the next section.

11.3.2 Multiple Comparisons

Which means are different from which other means? At this point we cannot say. However, inspection of the confidence intervals above suggest that the mean from the junk food group is larger than the other two groups. There are a number of different techniques for testing the differences between individual means following an ANOVA. These test procedures are called **multiple comparisons** tests. The specific technique depends to some extent on whether the investigator planned to compare certain groups of means or individual means with certain other means before the experiment was conducted. These are called planned comparisons. In other cases the investigator does not know before the experiment is conducted which means are to be compared to which other means, and in fact, may wish to compare all possible pairs of means with

Table 11.4   Generalized ANOVA Table

Source of Variation	SS	df	MS	F
Among-groups	SS_a	df_a	$\frac{SS_a}{df_a}$	$\frac{MS_a}{MS_w}$
Within-groups	SS_w	df_w	$\frac{SS_w}{df_w}$	
Total	SS_t	df_t		

Table 11.5   ANOVA Table for Mouse Diet Example

Source of Variation	SS	df	MS	F
Among-groups	65.973	2	32.98	41.81
Within-groups	21.301	27	0.789	
Total	87.274	29		

Table 11.6 *Minitab* **Results for Example 11.1**

One-way ANOVA: WeightGain versus Group

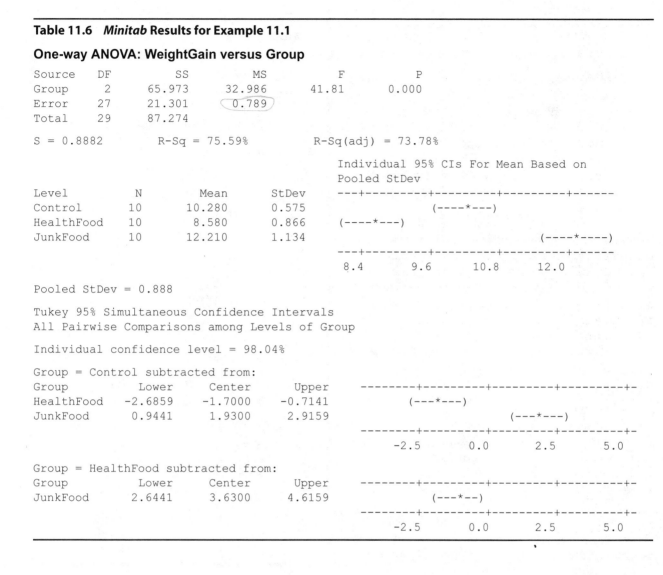

```
Source    DF       SS        MS         F        P
Group      2    65.973    32.986     41.81    0.000
Error     27    21.301     0.789
Total     29    87.274

S = 0.8882        R-Sq = 75.59%        R-Sq(adj) = 73.78%

                                   Individual 95% CIs For Mean Based on
                                   Pooled StDev
Level           N       Mean     StDev   ---+---------+---------+---------+------
Control        10     10.280     0.575                    (----*---)
HealthFood     10      8.580     0.866    (----*---)
JunkFood       10     12.210     1.134                                 (----*----)
                                         ---+---------+---------+---------+------
                                          8.4       9.6      10.8      12.0

Pooled StDev = 0.888

Tukey 95% Simultaneous Confidence Intervals
All Pairwise Comparisons among Levels of Group

Individual confidence level = 98.04%

Group = Control subtracted from:
Group          Lower     Center     Upper    --------+---------+---------+---------+-
HealthFood   -2.6859    -1.7000   -0.7141           (---*---)
JunkFood      0.9441     1.9300    2.9159                        (---*---)
                                             --------+---------+---------+---------+-
                                                  -2.5       0.0       2.5       5.0

Group = HealthFood subtracted from:
Group          Lower     Center     Upper    --------+---------+---------+---------+-
JunkFood      2.6441     3.6300    4.6159               (---*--)
                                             --------+---------+---------+---------+-
                                                  -2.5       0.0       2.5       5.0
```

each other. These are called **unplanned comparisons**. In this chapter we will deal with only one technique for unplanned comparisons, the extended Tukey test for multiple comparisons, because it is computationally relatively simple. There are a number of other such tests that one may use. Detailed descriptions are given in Sokal and Rohlf (2012). These procedures are typically listed as options with computer statistics packages in their ANOVA routines.

In example 11.1 we rejected H_0: $\mu_1 = \mu_2 = \mu_3$. The error mean square for this test was 0.789 with 27 df, and n for all three treatments was 10. We may now compute a critical value that the difference between any two means must equal or exceed to be considered significantly different from each other. This critical value (CV) for samples of equal size is given by

$$CV = q\left(\sqrt{\frac{MS_e}{n}}\right) \qquad \textbf{(11.4)}$$

where MS_e is the error mean square, n is the sample size (within each group), and q is the studentized range value, found in table A.7, "Critical Values of q for the Tukey Test." The value for q is found in table A.7. The top row of the table gives the number of groups (treatments), designated by a, and the left column gives the error degrees of freedom. The cell entry is the value of q. Since 27 df is not listed, we go to the next lower value listed (at 24 df), which is 3.53. Any error caused by doing this is a conservative one. (If a more accurate value for q is needed, it may be obtained by interpolation.) Thus, the critical value is

$$CV = 3.53\left(\sqrt{\frac{0.789}{10}}\right) = 0.9915$$

We now calculate the absolute difference between each pair of means.

Control – Junk food: $|10.28 - 12.21| = 1.93$
Control – Health food: $|10.28 - 8.58| = 1.70$
Junk food – Health food: $|12.25 - 8.58| = 3.67$

Since the differences between the various pairs of means are all larger than the CV, we may conclude that all of the means are significantly different from each other at the 0.05 level of significance. A diet of potato chips, Twinkies, and diet cola induces a greater rate of weight gain in laboratory mice than does regular mouse food, which in turn induces more weight gain than does granola and organically grown prune juice.

An alternative way of considering this same information is to compute confidence intervals for the difference between each pair of means (recall section 10.2), making a slight modification of the critical value. This is what is displayed by *Minitab* above. The fact that all the confidence intervals fail to include zero indicates that each of the means is different from one another.

11.3.3 Fixed-Effects ANOVA Using Survey Data

"Treatments" in ANOVA are not necessarily something that an experimenter "does" to groups of individuals, as shown in example 11.1 (also called a **manipulative experiment**). Rather, we may wish to determine if some measurement variable varies significantly among populations. Consider Example 11.2.

This is a fixed-effects model, since the habitats (treatments) were not chosen at random but were selected by the investigator. The term "treatment" in this situation does not have its usual meaning, since nothing was treated. Rather, treatment refers here to the specific locations from which the fish were sampled. Note that in this situation the investigator did not manipulate the situation beyond choosing the habitats to be sampled. Thus, we are using survey data rather than data from a manipulative experiment.

The calculations for a fixed-effects ANOVA using observational data are exactly the same as when using experimental data. The major difference is in the interpretation. In example 11.1 we can be fairly certain that different diets caused differences in weight gain in our experimental animals because each animal was randomly assigned to each treatment group. In situations like example 11.2, it would not be accurate to conclude that different lakes or streams cause differences in the size of sticklebacks (if such differences are revealed by the ANOVA) because individual fish were not randomly assigned to the various habitats; they were

Example 11.2
A Fixed-effects ANOVA Design Using Survey Data

Random samples of 20 sticklebacks (a small minnow) were each collected from three small lakes and three small streams in Michigan. We wish to know if these populations differ in total length (in millimeters) among these six habitats. The data are shown in table 11.7.

Table 11.7   Data for Example 11.2 Body Length (mm) of Sticklebacks. (Data are also available in digital appendix 4.)

Lake A	Lake B	Lake C	Stream A	Stream B	Stream C
31	36	28	47	47	38
32	30	38	48	37	36
34	32	31	50	41	48
34	37	32	42	38	43
35	35	29	44	32	42
30	32	38	34	45	31
33	32	40	41	42	40
32	37	36	40	40	45
37	39	43	44	43	42
33	28	34	47	40	49
36	32	32	39	39	39
30	31	39	47	45	30
32	35	31	43	41	42
39	40	36	40	39	39
30	36	28	38	32	38
29	31	39	32	48	35
42	32	32	41	32	49
39	27	38	45	45	40
37	35	29	42	41	43
29	31	32	37	38	42

already there. Thus, while we may detect differences among habitats, we should not conclude that the habitats directly cause the differences in the same sense that the diets in example 11.1 could be considered to be the cause of the differences in weight gain. If we detect differences which correspond to particular environmental features (such as, say, temperature), we could say that body length was associated with this feature. But a manipulative experiment is required to verify its cause.

If you worked your way through example 11.1 and tried your hand at a couple of the problems at the end of this chapter, you have developed a feeling about ANOVA by calculator, pencil, and paper. If you are like most people, you found the process tedious! Try example 11.2 using computer software. Table 11.8 shows that analysis done with *Minitab*. The analysis clearly shows that the sticklebacks differed in mean body length among the six popula-

Table 11.8 *Minitab* **Output for One-Way ANOVA on Stickleback Data (from table 11.7)**

One-way ANOVA: BodyLength versus Location

```
Source      DF          SS          MS          F           P
Location     5      1585.5       317.1       16.38       0.000
Error      114      2206.4        19.4
Total      119      3791.9

S = 4.399           R-Sq = 41.81%           R-Sq(adj) = 39.26%

                                            Individual 95% CIs For Mean Based on Pooled
                                            StDev
Level        N        Mean      StDev       +---------+---------+---------+---------
Lake A      20      33.700      3.672         (----*-----)
Lake B      20      33.400      3.485       (----*-----)
Lake C      20      34.250      4.459          (-----*----)
Stream A    20      42.050      4.685                             (----*-----)
Stream B    20      40.250      4.667                         (-----*-----)
Stream C    20      40.550      5.186                         (-----*----)
                                            +---------+---------+---------+---------
                                           31.5      35.0      38.5      42.0

Pooled StDev = 4.399
```

tions. The non-overlapping confidence intervals between the lake and stream populations suggest each habitat type forms a distinctive group. A follow-up Tukey test confirmed that these differences are real.

Following a multiple comparisons test, most users summarize the results with letter codes in either a table or a graph. For the stickleback example, figure 11.3 shows a bar graph, with identical letters indicating groups that are not significantly different from one another.

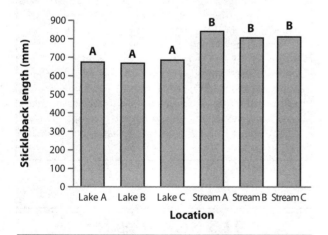

Figure 11.3 Bar graph of the Stickleback data from example 11.2. Matching letters above the bars indicate groups that are not significantly different from one another. For simplicity, standard error bars are omitted.

11.3.4 One-Way ANOVA Design with Random Effects

In examples 11.1 and 11.2 we were dealing with fixed-effects models (sometimes called model I ANOVA) because the treatments (diets, habitats) were selected by the investigator and we had no interest in generalizing the results to other diets or other habitats. In other words the questions asked were "Do these three diets differ with respect to weight gain induced in mice?" and "Is there a difference in the size of sticklebacks among these three lakes and three streams?" On occasion we wish to ask a somewhat different question, such as "Is fish size variable among lakes?" In such a case, we might select a number of different lakes at *random* from a large population of lakes and measure samples of fish from each lake. Here, we are not so interested in knowing about fish in particular lakes, but rather the size variation in this fish species across a large suite of lakes.

An ANOVA design using random effects is sometimes called a **random-effects (model II) ANOVA**. For the case just described, the null hypothesis is that fish are similarly variable across lakes as they are within lakes. Alternatively, the fish size shows greater variability among lakes than within lakes. If the null hypothesis is rejected, we do not conduct a multiple-comparisons test in this type of ANOVA, since we have no interest in knowing which particular lake is different from which others. We only want to know if the variability is greater or not. However, we might want to know how much of the variation in body size is due to differences *among* populations versus differences among individuals *within* populations. This basically examines a breakdown of variance components. Such questions are very common in the field of quantitative genetics. Exploring this analysis is beyond the scope of this text, so interested readers should consult another text, such as Sokal and Rohlf (2012), for a complete discussion.

11.4 Testing the Assumptions of ANOVA

As we have seen (section 11.2), for ANOVA to work correctly, several assumptions must be met. A proper experimental design should assure that subjects are selected at random. This will also assure independence of errors. The errors are "unexplained noise" which is not accounted by the treatments we impose. These errors are sometimes called **residuals**, the deviation of a measurement from its group mean. Recall that we earlier estimated the within-groups variance, also known as the error variance. This is simply the variance of these residuals.

When individuals from a population are randomly assigned to treatments in a carefully controlled experiment, and assuming that the treatments do not in some way affect variability of a measured response, homogeneity of within-groups variance usually results. In example 11.1, which is a fairly typical application of a completely randomized design, the use of genetically similar individuals (inbred mice) randomized among treatment groups should result in homogeneity of error variances. Consider example 11.2, concerning the length of sticklebacks from six different habitats. In this case individuals could not be randomly assigned to the different habitats, since they were already there, and it would be unwise to assume that the variances among these six populations are equal. Descriptive statistics from this example are presented below.

Table 11.9 Group Means and Variances for Sticklebacks from Example 11.2

	$\bar{x}$	s^2
Lake A	33.70	13.484
Lake B	33.40	12.147
Lake C	34.25	19.882
Stream A	42.05	21.945
Stream B	40.25	21.776
Stream C	40.55	26.892

Typically, for ANOVA, we check two assumptions, normality (of residuals) and homogeneity of (error) variance. In section 8.4, we saw how normality can be checked using a graphic technique. Here we will briefly see how these assumptions are checked for ANOVA using statistical software. Most software packages that do ANOVA have routines for checking these assumptions. For example, using *Minitab*, we checked these assumptions for the stickleback data and have displayed the results in figures 11.4 and 11.5. To check the normality assumption, we first had to store the residuals while doing the ANOVA; these residuals were then stored as a new column of data in the worksheet.

On the normality plot, notice that the points fall close to a straight line, indicating no significant departure from normality. This result was further confirmed by a goodness of fit test, the Anderson-Darling test ("AD"), which checks how closely the points (here the residuals) follow a normal distribution. The high p-value ($p = 0.491$) indicates that there is no cause to reject the null hypothesis of normality. Thus, we safely assume normality.

Figure 11.5 illustrates *Minitab* results from a test for homogeneity of error variances. Although the sample variances in table 11.9 showed some differences among groups, there is no evidence that the population variances are different. Two different analyses, Bartlett's Test and Levine's Test, each give high p-values for the null hypothesis of equal error variances. Thus we are safe to assume homogeneity of variance for the stickleback data.

Caution

There are a number of important assumptions of ANOVA, one of which is that the within-groups (error) variances of the various groups are all the same. ANOVA is said to be robust to the assumption of normality, meaning that some departure from this assumption is not too serious, particularly when large samples are involved. But ANOVA is very sensitive to the homogeneity of variances assumption. When this assumption fails, ANOVA may give us the wrong answer and so should not be used.

11.5 Remedies for Failed Assumptions

There are situations in which the data collected in an experiment do not meet the normality and equal variance assumptions, even though the variable in question consists of continuous measurements. When these problems exist, a mathematical transformation might correct the problem. Transformations are discussed below. When these assumptions are not satisfied by the transformed data or when variable consists of ranks (ordinal data), one should use a nonparametric test. For the one-way ANOVA design, the nonparametric analogue is the Kruskal-Wallis test (section 11.5.2). In no case should a parametric ANOVA be used with ordinal data! As mentioned earlier, the error introduced by using a discrete measurement variable is not large provided that the number of values the variable can

assume is fairly large. This is assured when sample sizes are large.

11.5.1 Transformations in ANOVA

A **transformation** involves performing the same mathematical operation on each observation in a set of data. (This is easily done in a spreadsheet or with statistical software.) In the following sections, some common transformations and their uses are discussed. For convenience, original observations are designated by the usual x, and transformed observations are designated by x'.

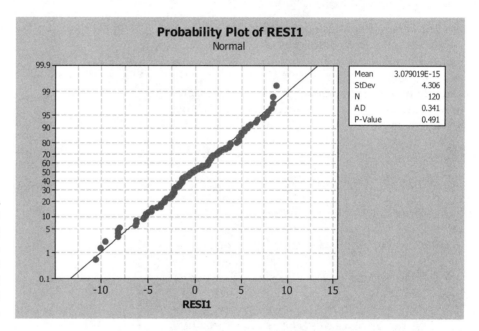

Figure 11.4 *Minitab* output for checking normality assumption. "RESI1" is the set of residuals generated from ANOVA on the stickleback data (example 11.2).

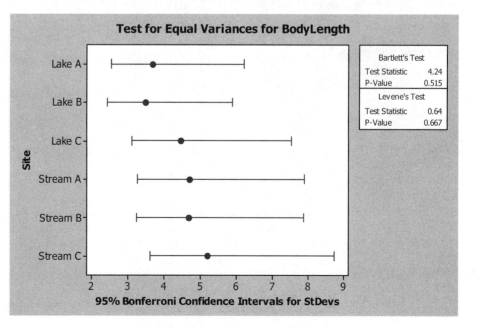

Figure 11.5 *Minitab* output for checking homogeneity of variance assumption for the stickleback data (example 11.2).

The Logarithmic Transformation

The logarithmic transformation is often useful in ANOVA and in linear regression (chapter 14) where the error variances are not equal. The log transformation is conducted by taking the logarithm of each observation, or

$$x' = \log(x) \qquad (11.5)$$

Natural logarithms or base 10 logarithms are commonly used. When there are zeros in the data, a constant may be added to each observation before taking the log. For example,

$$x' = \log(x + 1) \qquad (11.6)$$

since log (0) is undefined.

The Square Root Transformation

When data are in the form of counts (integer values), a square root transformation often helps improve normality of the distribution and/or equality of variances among groups. This transformation involves taking the square root of each observation, or

$$x' = \sqrt{x} \qquad (11.7)$$

The Arcsine Transformation

Percentages and proportions tend to be not quite normally distributed, although the underlying variable with which they deal might be. Since proportions range from 0 to 1 and percentages range from 0 to 100, the "tails" of such a distribution tend to be compressed. This situation is noticeable when many of the observations fall below 30% (0.3) or above 70% (0.7). This problem can often be rectified by an arcsine (sine^{-1}) transformation, which, for proportions, is

$$x' = \text{arcsine } x \qquad (11.8)$$

or for percentages is

$$x' = \text{arcsine}\left(\frac{x}{100}\right) \qquad (11.9)$$

Following transformation, transformed values are checked to see that they meet ANOVA assumptions. If all is OK, then the ANOVA is run on the transformed values. If the null hypothesis is rejected and we conclude that not all means are equal for the transformed value, we can further conclude that the means for the original values are not all equal. We can also use similar logic when interpreting the results from a multiple comparisons test. For descriptive purposes, we can back transform the descriptive statistics. (For example, if the square root function was used for the transformation, the square function is used for back transformation.)

11.5.2 Nonparametric Alternative to One-Way ANOVA: The Kruskal-Wallis Test

When data consist of ranks or when data consist of measurements but the assumptions for ANOVA are not otherwise met, one may use a nonparametric test to compare three or more groups. These tests are often called nonparametric ANOVA tests, but since we do not measure variance in nonparametric tests, the name is a bit misleading. In effect, these tests tell us if it is reasonable to assume that three or more samples could have been drawn from identical populations.

The Kruskal-Wallis test is the nonparametric counterpart of the one-way ANOVA. It is used to test the null hypothesis that three or more independent samples were drawn from identical populations.

Example 11.3
The Kruskal-Wallis Test

The time (in seconds) that males of a certain species of grasshopper remained mounted on females during mating was determined. 10 males each were randomly assigned to one of four treatments (conditions). The data are as follows.

Treatment A	Treatment B	Treatment C	Treatment D
9	30	3,900	5
3	30	10,800	9
10	480	28,900	20
200	900	3,600	180
1	2	200	15
2	1	120	20
21	5,400	500	2
720	1,500	600	17
1,500	480	1,980	30
60	3	160	8

Modified data from R. Bland

The means, medians (m), and variances for the 4 treatments (conditions) are

	Treatment A	Treatment B	Treatment C	Treatment D
$\bar{x}$	252.6	882.6	5076.0	30.6
m	15.5	255	1290	16
s^2	242,064	2,765,569	80,766,169	2,830

Although the experiment is properly designed for a completely randomized one-way ANOVA, and the data consist of measurements (time), there are a couple of assumptions of the parametric test that these data do not satisfy. First, the variable does not seem to be even approximately normally distributed, as indicated by the large differences in the means and the medians of each group. As you will recall from chapter 8, a normal distribution is symmetrical about the mean and in such distributions, the mean and median should be about the same value (fig. 4.1). Secondly, the within-groups variances are clearly not equal. (And a check of the homogeneity assumption with *Minitab* confirms this suspicion.) Accordingly, we will use the Kruskal-Wallis test.

Assumptions of the Test
1. The sampled populations have similar shape to their distributions.
2. The samples represent random samples which are independent.
3. Data consist of ranks or measurements.

To conduct the Kruskal-Wallis test, all of the observations in all of the groups are ranked with respect to each other, with the lowest value in any of the samples receiving the rank of one. Tied values receive the average rank that they would have if they were not tied, as in previous examples. The ranked data and the sum of ranks for each group for the example are as follows.

Group A	Group B	Group C	Group D
10.5	19	37	8
6.5	19	39	10.5
12	27.5	40	15.5
25.5	32	36	24
1.5	4	25.5	13
4	1.5	22	15.5
17	38	29	4
31	33.5	30	14
33.5	27.5	35	19
21	6.5	23	9
ΣR 162.5	208.5	316.5	132.5

The test statistic used in the Kruskal-Wallis test is designated as H, which, when the null hypothesis that all of the samples were drawn from identical populations is true, has a chi-square distribution. H is computed by:

$$H = \frac{12}{n_t(n_t+1)}\left(\Sigma \frac{(\Sigma R_i)^2}{n_i}\right) - 3(n_t+1) \quad \textbf{(11.10)}$$

where n_i = number of observations in a group, ΣR_i = the sum of ranks in a group, and n_t = the total number of the observations in all of the groups.

For the example

$$H = \frac{12}{40(40+1)}\left(\frac{(162.5)^2}{10} + \frac{(208.5)^2}{10} + \frac{(316.5)^2}{10} + \frac{(132.5)^2}{10}\right) - 3(40+1)$$

and after crunching the numbers:

$$H = 14.273$$

H has a chi-square distribution with degrees of freedom equal to the number of groups minus one. Consulting table A.3, we find that our calculated value of H is greater than the critical value of chi-square with 3 df at the 0.05 level (7.81). The p-value is $.001 < p < .01$. We thus may reject H_0 and conclude that the four samples could not have been drawn from identical populations.

Doing a Kruskal-Wallis test with paper and calculator is quite tedious. Doing the ranking is considerably easier using a spreadsheet (hint: put all the measurements in one column, group labels in a second column, sort by measurements, and put the numbers $\{1,2,3,..., n_t\}$ in a third column). This is a good exercise once or twice, but then we can rely on statistical packages to do these statistics on a routine basis. Results from *Minitab* are displayed below.

Table 11.10 *Minitab* **Kruskal-Wallis Test Results for the Grasshopper Mounting Data (example 11.3)**

Kruskal-Wallis Test: TimeMounted versus Condition

```
Kruskal-Wallis Test on TimeMounted

Condition     N    Median    Ave Rank       Z
Treatment A  10     15.50        16.3   -1.33
Treatment B  10    255.00        20.9    0.11
Treatment C  10   1290.00        31.7    3.48
Treatment D  10     16.00        13.3   -2.26
Overall      40                  20.5

H = 14.27   DF = 3   P = 0.003
H = 14.29   DF = 3   P = 0.003  (adjusted for
ties)
```

Minitab reports the same statistic as we calculated above and the exact p-value shown falls within the range we determined from the chi square table.

In this chapter we examined the one-way Analysis of Variance (ANOVA) layout and the multiple comparisons test, which follows rejecting the null hypothesis of a fixed-effects model. We also explored how to check the assumptions of ANOVA, and used two approaches for dealing with data which fail to meet these assumptions. In the following chapter, we'll examine two other ANOVA layouts which are widely used, the factorial and randomized block designs.

Key Terms

among-groups variance (treatment variance)
Analysis of Variance (ANOVA)
fixed-effects (model I) ANOVA
homogeneity of variance
Kruskal-Wallis test
multiple comparisons
random-effects (model II) ANOVA
residuals
transformation
Tukey test
within-groups variance (error variance)

Exercises

For each of the following problems, state the null and alternative hypotheses, being specific to the particular problem of interest. Use the appropriate computations to calculate the necessary statistics for an ANOVA table. Determine the p-value, decision on the validity of the null hypothesis, and conclusion about the original problem of interest. If appropriate, conduct a multiple-comparisons test and conclusions from that test.

Exercises 11.1 through 11.4 use a completely randomized one-way ANOVA design. You may assume a normal distribution and equality of within-groups variances.

11.1 Fifteen tobacco plants of the same age and genetic strain were randomly assigned to 3 groups of 5 plants each. One group was untreated, one was infected with tobacco mosaic virus (TMV), and one was infected with tobacco ringspot virus (TRSV). After one week the activity of odiphenol oxidase was determined in each plant. Does infection by either virus affect the activity of this enzyme?

Enzyme Activity ($\mu l\ O_2$/mg protein/min)

Control	TMV-Infected	TRSV-Infected
1.47	2.44	2.87
1.62	2.31	3.05
1.06	1.98	2.36
0.89	2.76	3.21
1.67	2.39	3.00

11.2 Eighteen freshwater clams were randomly assigned to 3 groups of 6 each. One group was placed in the pond water from which the clams were collected, one group was placed in deionized water, and one group was placed in a solution of 0.5 mM sodium sulfate. At the end of a specified time period, blood potassium levels were determined. Do the treatments affect blood potassium levels (μM K^+)?

Pond Water	Deionized Water	Sodium Sulfate
.518	.318	.393
.523	.342	.415
.495	.301	.351
.502	.390	.390
.525	.327	.385
.490	.320	.397

Data from J. Schiede

11.3 Cellulase activity (units/mg protein) in 4 genetic variants of a fungus species was measured in 5 randomly selected cultures of each variant. Is there a difference in activity of this enzyme in the four variants? Note that the variants are chosen at random; thus we are not interested in specific means, only in knowing if added variation is introduced.

Variant A	Variant B	Variant C	Variant D
10	20	15	5
12	21	18	3
9	19	13	6
11	23	14	6
10	18	12	4

11.4 Fifteen juniper pythons of similar size and age were randomly assigned to 3 groups. One group was treated with drug A, one group with drug B, and the third group was not treated. Their systolic blood pressure was measured 24 hours after administration of the treatments. Does either drug affect blood pressure? Does one have more or less of an effect than the other?

Drug A	Drug B	Untreated
118	105	130
120	110	135
125	98	132
119	106	128
121	105	130

Exercises 11.5 through 11.8 are appropriate for the Kruskal-Wallis test.

11.5 The aggressiveness of 15 randomly selected female gully cats was measured under 3 different conditions. Aggressiveness was measured on a rather subjective scale of 0 to 100 and should be regarded as an ordinal measurement. We wish to know if there is a difference in aggressiveness among the three conditions. The results were as follows.

Condition A	Condition B	Condition C
5	20	60
10	15	50
20	20	70
8	25	25
11	27	90

11.6 Fifteen inbred laboratory rats were randomly assigned to 3 groups of 5 each. Each rat was inoculated with *Staphylococcus mucans*, a bacterium involved in the process of tooth decay. One group (control) was given no additional treatment. A second group was given drug A, and a third group was given drug B. At the end of six weeks, the extent of tooth decay in the rats was evaluated as % of tooth decay. This is a fairly subjective measurement based on the combined evaluation of 3 observers and should probably be regarded as an ordinal measurement. We wish to know if there is a difference among the 3 groups.

Control	Drug A	Drug B
87	63	45
76	70	60
65	87	43
81	92	56
75	70	60

11.7 Three groups of eggs of a tropical frog were exposed to 3 concentrations of benzene: zero benzene, 50 ppm benzene, and water saturated with benzene. After hatching, the brain size of the tadpoles was measured. We wish to determine if exposure to benzene affects brain size. We do not know if this variable follows ANOVA assumptions, nor is it possible to check with a sample as small as this. The results were as follows.

Zero	50 ppm	Saturated
81	88	111
72	89	109
68	92	133
87	107	
	101	
	91	

Data from J. Martin

11.8 Twenty-four rug rats were randomly assigned to 4 groups of 6 each. The groups were subjected to different levels of stress (high, moderate, low, and none) induced by occasional, unexpected loud noises. After one week, density of blood lymphocytes (cells $\times 10^6$/ml) were measured. Prior studies indicated that the homogeneous assumption was violated and could not be remedied.

High	Moderate	Low	None
2.9	3.7	5.2	6.8
1.8	4.6	6.2	7.9
2.1	4.2	5.7	7.1
1.5	2.8	4.6	6.5
0.9	3.2	5.1	6.3
2.8	4.5	4.9	7.6

Use statistical software to solve exercises 11.9 through 11.14. First check the assumptions of ANOVA. If the assumptions are violated, use the appropriate nonparametric test.

11.9 Random samples of *Daphnia* were collected from the zooplankton of 5 randomly selected lakes and their selenium content was determined. We wish to know if there is a difference among lakes with respect to selenium content in these *Daphnia* (i.e., is there a significant "lake effect"?).

Lake A	Lake B	Lake C	Lake D	Lake E
23	34	15	18	25
30	42	18	15	20
28	39	12	9	22
32	40	10	12	18
35	38	8	10	30
27	41	16	17	22
30	40	20	10	20
32	39	19	12	19

11.10 Thirty pea seeds were randomly assigned to 3 groups of 10. One group was germinated in the presence of chemical A, a second group with chemical B, and the third group served as a control. After five days, the length (in millimeters) of the primary root was measured. We wish to know if either or both of these chemicals affects root growth, and if so, if one is more effective than the other.

Chemical A	Chemical B	Control
115	120	82
103	125	97
98	122	105
121	100	90
130	90	102
107	128	98
106	121	105
120	115	89
100	130	100
125	120	90

11.11 Thirty chickens of a highly inbred strain were randomly assigned to 3 groups of 10. One group was injected with 100 μg norepineph-rine per kilogram of body weight, a second group was injected with 100 μg epinephrine per kilogram of body weight, and a third group served as the control. SRBC plaque forming cells per 10^6 spleen cells were measured after a specified time interval. Is there a difference among the three groups?

Norepinephrine	Epinephrine	Control
15	70	535
155	45	370
110	95	420
90	95	315
35	70	485
100	315	230
30	140	370
40	260	320
75	230	335
105	400	475

11.12 Testosterone levels in mature roosters of three strains of chickens were measured, and the following results were obtained. Is there a difference among strains?

Strain A	Strain B	Strain C
439	102	107
568	115	99
134	98	102
897	126	105
229	115	89
329	120	110

11.13 The antibody response in three groups of diabetic mice (normal, alloxan diabetic, and alloxan diabetic treated with insulin). Is there a difference in the amount of antibodies among the three groups? Before testing the main hypothesis, check the ANOVA assumptions of homogeneous variance and normality. (If the data fail these assumptions, first try transforming the data to see if the data can be corrected; if the data still fail one or more assumptions, use the appropriate nonparametric test.)

Normal	Alloxan	Alloxan+insulin
156	391	82
282	46	100
197	469	98
297	86	150
116	174	243
127	133	68
119	13	228
29	499	131
253	168	73
122	62	18
349	127	20
110	276	100
143	176	72
64	146	133
26	108	465
86	276	40
122	50	46
455	73	34
655	44	14

Data from Hand et al. 1984

11.14 Fifteen turkey hens were randomly assigned to 3 groups of 5. One group was given diet A, the second group diet B, and the third group diet C. We wish to know if there is a difference in the weight of eggs produced by the birds on these diets, and if so, which diet results in the largest eggs. The data are the mean weights of 10 eggs from each bird.

Diet A	Diet B	Diet C
124	98	116
118	100	97
120	95	100
127	102	89
115	105	98

More ANOVA
Factorial Design and Randomized Block

In chapter 11, we explored one-way Analysis of Variance (ANOVA) and its nonparametric analogue, the Kruskal-Wallis test. The one-way design looks at the effect of changing a single categorical factor ("treatment" or "factor") on a measurement of interest ("response"). The design is also completely randomized, in that subjects were randomly chosen to receive particular treatments. ANOVA can be extended to many other experimental designs. In this chapter, we'll briefly examine two of them: the randomized block design and the factorial design. Since the hand calculations are quite tedious, we'll use computer software to do all the number crunching. Our goals are to make sense of the statistical results and to be aware of certain design principles to guide us past any pitfalls.

12.1 The Randomized Block Design

Recall example 11.1, where we fed groups of mice one of three different diets and then measured their weight gain. We deliberately chose a strain of inbred mice, to eliminate effects of genetic differences, and had a clear record of their history in the lab, so we could pick even-aged individuals for the experiment.

Suppose that our interest was in a wild population of white-footed deer mice rather than in highly inbred laboratory mice. We might expect that genetic variability would be fairly high in such a population. Furthermore, we would not know their ages. Using the completely randomized design (section 11.3), the additional variance in growth rates due to genetic and age differences among the experimental units would add to the error variance. This larger error variance would have the effect of reducing the variance ratio and making it more difficult to reject the null hypothesis (i.e., reduced power of the test).

The **randomized block design** provides a way to deal with this problem. In this design individuals are "blocked" (grouped) according to the characteristic whose variance we wish to identify and "partition out." In the case of our experiment with wild mice we would group individuals according to litters, since it is likely that members of the same litter are genetically similar and of the same age. Note that each individual observation (each mouse) is classified according to two criteria (diet and litter) and that only one individual occupies each possible combination of diet and litter. Thus, this particular design is sometimes called a two-way ANOVA without replication. It is a mixed effects model, where diet is a fixed factor and litter is a random factor.

As in the completely randomized one-way ANOVA design, the null hypothesis in a randomized block design is that the treatment means are equal. We are also interested in knowing whether block (in this case, litter) is also an important source of variation. The null hypothesis for this second question is that block effect is unimportant.

Example 12.1
A Randomized Block Design

Ten litters of white-footed deer mice of approximately the same age were selected. One member of each litter was randomly assigned to one of the three treatment groups (the diets of example 11.1), and their weight gain was determined as before. The results are given in table 12.1.

Table 12.1   Weight Gain by White-footed Deer Mice Fed Mouse Food (A), Junk Food (B), and Health Food (C)

	Litter Number	Group A	Group B	Group C
		Treatments		
	1	11.8	13.6	9.2
	2	12.0	14.4	9.6
	3	10.7	12.8	8.6
	4	9.1	13.0	8.5
Blocks	5	12.1	13.4	9.8
	6	9.8	10.9	10.0
	7	10.5	13.6	9.2
	8	10.5	11.9	8.8
	9	9.0	10.5	6.9
	10	11.2	13.8	10.1
	$\bar{x}$	10.67	12.79	9.07

Several features about this design require attention. Treatments, in this case, are the different diets and represent the factor in which we have our primary interest. Treatments are customarily shown as separate columns. Blocks, which are the individual litters, represent the source of variation we regard as extraneous and we wish to remove (or partition out) from what would otherwise be a part of the error variance. Blocks are shown as rows. Each treatment contains one and only one member of each block. In the example we would randomly select one member of each litter for each treatment. The total number of observations (n_t) in such a design is therefore the number of treatments (diets) × the number of blocks (litters). So, in this example, the total numbers of observations is 3 × 10 = 30.

The calculations for the randomized block design are similar to those for the completely randomized design, except that they are a bit more extensive in order to partition the variance into more components. For this design, we wish to partition the total sum of squares into a sum of squares associated with the treatments, a sum of squares associated with the blocks, and an error sum of squares. We will skip any further calculations of sums of squares, leaving that to *Minitab* (table 12.2).

The degrees of freedom are determined by the number of different treatments and the number of different blocks. For the current example, we have three diets and 10 litters. Total degrees of freedom are $df_t = n_t - 1$. For the example, $df_t = 30 - 1 = 29$. Column (treatment) degrees of freedom are $df_c =$ columns − 1. For the example, $df_c = 3 - 1 = 2$. Row (block) degrees of freedom are $df_r =$ rows − 1. For the example, $df_r = 10 - 1 = 9$. Error degrees of freedom are $df_e = df_t - df_c - df_r$. For the example, $df_e = 29 - 2 - 9 = 18$. Table 12.2 shows the ANOVA table for this example, as generated by *Minitab*.

Table 12.2   *Minitab* Results for Example 11.1, Effects of Diet and Litter on Weight Gain in Wild Mice. This example uses a randomized block design, which *Minitab* runs using the two-way ANOVA routine.

Two-way ANOVA: WeightGain versus Diet, Litter

```
Source   DF        SS       MS        F       P
Diet      2    69.643   34.8213   73.25   0.000
Litter    9    25.934    2.8815    6.06   0.001
Error    18     8.557    0.4754
Total    29   104.134

S = 0.6895 R-Sq = 91.78% R-Sq(adj) = 86.76%
```

The F statistic (variance ratio) for diet treatments (73.25) is found by dividing the treatment mean square by the error mean square. (Remember that "mean square" refers to the same thing as a variance.) The critical value of F is found in table A.6 for $\alpha = 0.05$ at 2 and 18 degrees of freedom. Since 18 is not tabulated, we'll use the values for 2 and 20 df ($F_{crit} = 3.49$) and for 2 and 15 df ($F_{crit} = 3.68$). Since our calculated F value is much larger than both those numbers, we reject H_0 and conclude that diet has a significant effect on weight gain in white-footed deer mice. It is also possible to test for significant differences among blocks. For this hypothesis, $F = 6.06$, which we compare to the critical value at 9 and 18 degrees of freedom. $F_{.05, 9, 20} = 2.39$ and $F_{.05, 9, 15} = 2.59$. Again, since the F statistic (6.06) is much larger than these numbers, the second hypothesis is also rejected. Substantial variation was factored out with the different litters. This provides guidance for future experiments that using a block design is the best way to go when working with these animals.

A very common use of the randomized block design is when one desires to make three or more repeated measurements on the same individual. Thus, it may be used in much the same way that the

Example 12.2
A Randomized Block Design Using Repeated Measures

The convict cichlid is one of several fish species that exhibits biparental care of eggs and fry. An experiment was conducted to determine if male convict cichlids might spend more time in direct offspring care with fry than with eggs. Accordingly, the time eight males spent in this activity (in seconds) during a 15 minute observation period was determined for five consecutive days. The eggs hatched after two days and for several days after that the fry were attached to the substrate by sticky "pads" on their heads. The results are given in table 12.3.

Table 12.3 Brooding Time (sec/15 min) by Male Convict Cichlids

Male Number	Day 1	Day 2	Day 3	Day 4	Day 5
1	11.9	2.2	57.9	259.5	200.4
2	42.7	60.7	71.2	163.3	228.1
3	15.8	14.8	311.3	283.9	436.3
4	191.2	148.8	437.8	319.2	462.6
5	3.5	187.3	281.4	410.4	373.7
6	23.7	0.0	98.6	185.7	106.8
7	0.0	0.0	102.4	400.7	386.9
8	33.5	107.5	193.5	317.8	337.3

Data from D. Dickens

paired t test is used in a before-after study (section 9.6), except that more than two measurements are involved. (The paired t test is, in fact, a special case of the randomized block design.) When used in this way, the randomized block design is sometimes called "**repeated measures ANOVA.**"

In this case, days are treatments (shown as columns) and individual males are blocks (shown as rows). The randomized block design is appropriate in this case because we wish to repeatedly measure the same males, and we wish to partition out the individual variation among males. You will note that this variation among males is considerable in this example and might well be large enough to mask any treatment effect if it were a part of the error variance. The computer solution for this example is given below in table 12.4. The results lead us to conclude that we may reject the first null hypothesis (no effect of offspring age). Treatment means are clearly quite different, as shown by the very low p-value ($p < 0.001$). Inspection of the means suggests that days 1 and 2 were not significantly different from each other but were significantly lower than days 3, 4, and 5. (This can be confirmed with a multiple comparisons test.) We thus conclude that males spend less time brooding eggs than brooding fry. The second null hypothesis is also rejected, supporting our observation that different males vary a lot in their time devoted to offspring. (Fishes are not too different from humans in this regard!)

Table 12.4 *Minitab* Results for Example 12.2, Brooding Time by Age of Young (Treatment) and Individual Male (Block). This example uses the same two-way ANOVA procedure to address a repeated-measures design.

Two-way ANOVA: BroodTime versus YoungAge, Male

```
Source     DF        SS        MS        F        P
YoungAge    4    513610    128402    24.24    0.000
Male        7    215656     30808     5.82    0.000
Error      28    148325      5297
Total      39    877590

S = 72.78        R-Sq = 83.10%        R-Sq(adj) = 76.46%

                          Individual 95% CIs For Mean Based on
                          Pooled StDev
YoungAge       Mean     -+---------+---------+---------+--------
Day 1        40.288     (----*----)
Day 2        65.163       (-----*----)
Day 3       194.263                      (----*-----)
Day 4        92.563                          (----*-----)
Day 5       316.513                             (-----*----)
                        -+---------+---------+---------+--------
                         0        100       200       300
```

12.2 The Friedman Test

The Friedman two-way ANOVA tests whether three or more related samples could have been drawn from identical populations. Thus, the Friedman two-way ANOVA is analogous to the randomized block design. The Friedman test is a nonparametric procedure, used when the assumptions of ANOVA (section 11.4) are not met.

To conduct the Friedman test, the data are arranged in a two-way table of k columns, representing the treatments, and n rows, representing the blocks. We will examine a dataset, use *Minitab* to do the hard part, then interpret these results. Essentially, *Minitab* will do an analysis with ranks and sum them in different ways to generate a statistic which can be compared to a chi-square distribution. The logic of this test is that if there were no difference among the treatments (columns), the sums of the ranks of the columns would be approximately equal. This is because any rank would be as likely to occur in any column as in any other. See example 12.3.

When n and k are not too small, the calculated test statistic has a chi-square distribution. We may therefore consult table A.3 to determine if the differences among the eight treatments are significant. The degrees of freedom in the Friedman test are $k - 1$, which, for the example, is $8 - 1 = 7$. The critical value of chi-square at alpha = 0.05 with 7 degrees of freedom is 14.07. Note that this test does not allow us to determine if the block effect is important. See table 12.6.

Since the calculated value of the test statistic ($S = 17.44$) is greater than 14.07, we may reject the null hypothesis that there is no significant difference between the treatments. The low p-value (0.015) is evidence that the treatment effect is discernible. Since,

---

Example 12.3
The Friedman Two-Way ANOVA

Do neonatal garter snakes exhibit a decremental response (habituation) to a repeated overhead stimulus? Six snakes were chosen at random and were placed in the experimental chamber and allowed to acclimate for approximately 30 minutes. An overhead rapidly moving object was then presented at 10-second intervals, and the reaction of the snake was noted. Reactions were scored from 3 (a rapid retreat from the stimulus), through 0, which indicated no response.

These are ordinal data, so a nonparametric test must be used. Since each animal was tested repeatedly, the observations are related and not independent. The data of the experiment are shown in table 12.5.

Table 12.5   Data for Example 12.3: Reaction Scores of Snakes

Snake Number (Block)	Interval (Treatment)							
	1	2	3	4	5	6	7	8
1	3	3	2	2	0	1	0	0
2	3	2	0	2	0	1	0	0
3	2	0	0	0	3	0	0	0
4	2	2	0	0	0	0	0	0
5	3	0	2	0	2	0	0	0
6	2	2	0	1	1	1	1	0

Data from R. Hampton and J. Gillingham

Table 12.6   Results of Friedman's Test on Reaction Scores of Snakes (example 11.5)

Friedman Test: ReactionScore versus Interval blocked by Snake

```
S = 17.44 DF = 7 P = 0.015
S = 22.96 DF = 7 P = 0.002 (adjusted for ties)

                          Sum
                    Est    of
Interval   N     Median   Ranks
I-1        6      2.328    45.5
I-2        6      1.703    35.5
I-3        6      0.328    23.0
I-4        6      0.578    26.5
I-5        6      0.453    27.0
I-6        6      0.453    23.5
I-7        6      0.266    19.0
I-8        6      0.016    16.0

Grand median = 0.766
```

in this case, "treatments" are the successive presentations of the test stimulus at 10-second intervals, we further conclude that neonatal garter snakes exhibit a decremental response (i.e., they habituate) to the stimulus.

12.3 The Factorial Design

In many situations two or more factors (treatments) interact with each other to produce effects beyond the sum of the effects of the two acting alone. In other words, factors (treatments) may interact either synergistically or antagonistically. For example, there are a number of medications that should not be taken together. Either drug taken alone might produce its desired effect, but when taken together they interact in some harmful way.

When **interaction** among two or more factors is suspected, the **factorial design** is appropriate. Note that the factorial design is similar in layout to the randomized block design, except that both factors represent treatments in which we have an interest and that each combination of treatments (cell) consists of replicated observations. Quite often both treatments are fixed effects. Since each individual is classified according to two criteria—the two main treatments—and since there are several individuals (replicates) within each combination of the two treatments, this design is sometimes called two-way ANOVA with replication.

Example 12.4
A Factorial Design ANOVA

Consider another mouse diet experiment similar to the two examples used earlier to illustrate the completely randomized (example 11.1) and the randomized block designs (example 12.1). In this experiment, however, we wish to determine the effect of diet, the effect of stress, and the interaction of these two factors, if any, on weight gain. We could conduct two separate experiments—one on diet and one on stress—but this would give us no information on their possible interaction.

For this experiment, we select 32 highly inbred mice of the same age and sex (to minimize the error variance) and assign them at random to four groups. One group will receive the potato chip-Twinkie-cola diet (junk food diet) and will listen to rap music eight hours each day (high stress). Another group will have the same diet but will listen to Baroque music for eight hours each day (low

stress). A third group will receive regular mouse food (control) and will listen to rap music (high stress). The fourth group will receive the control diet and will listen to Baroque music. (The musical bias of the authors is evident here!) All four groups will be housed under identical conditions. Thus, the two variables under study are arranged in all four possible combinations. Each combination of treatments will have eight replicate mice ($n = 8$). This is an example of a 2×2 factorial design.

We have an interest in two main effects (or treatments), diet and stress, and a possible interaction effect, which we wish to detect if it is present. We test three null hypotheses in a situation like this.

1. Effect of stress
   $H_0: \mu_L = \mu_H$ vs. $H_a: \mu_L \neq \mu_H$
2. Effect of diet
   $H_0: \mu_C = \mu_J$ vs. $H_a: \mu_C \neq \mu_J$
3. Interaction effect
   H_0: No Interaction vs. H_a: Interaction

Data from this experiment are shown in table 12.7. The 32 measurements of weight gain (response variable) are arranged in four groups, representing the four combinations of treatments. We also show the cell means, which will be illustrated later. We will not go through the cumbersome calculations, but will use *Minitab* to generate the necessary statistics.

Table 12.7 Effect of Diet and Stress Level on Weight Gain in Mice. Weight gain is measured as mg increased in one week.

Diet	Stress Level Low	Stress Level High
Junk food	132	157
	128	143
	142	162
	131	150
	135	149
	120	140
	139	159
	133	158
	mean **132.5**	mean **152.25**
Control	120	130
	131	142
	122	131
	129	124
	120	124
	119	131
	134	143
	123	131
	mean **124.75**	mean **132**

The data for ANOVA may also be arranged in another way, as shown below in table 12.8. This data layout is the structure commonly used by statistical packages, such as *Minitab*. In this layout, all the measurements of the response variable are listed in one column and codes are placed in two other columns to indicate which treatment condition (level) is associated with each measurement. For the present example, the data matrix has 32 rows, for the 32 total mice in the experiment.

Table 12.8   Effect of Diet and Stress Level on Weight Gain in Mice. Data are arranged in a layout that most statistical packages accept. Diets: J—junk food, C—control; Stress levels: L—low, H—high.

Weight Gain	Diet	Stress	Weight Gain	Diet	Stress
132	J	L	157	J	H
128	J	L	143	J	H
142	J	L	162	J	H
131	J	L	150	J	H
135	J	L	149	J	H
120	J	L	140	J	H
139	J	L	159	J	H
133	J	L	158	J	H
120	C	L	130	C	H
131	C	L	142	C	H
122	C	L	131	C	H
129	C	L	124	C	H
120	C	L	124	C	H
119	C	L	131	C	H
134	C	L	143	C	H
123	C	L	131	C	H

The degrees of freedom for each main effect is the number of groups minus one. For the example, df_{diet} and df_{stress} are both $2 - 1 = 1$. The degrees of freedom for interaction is degrees of freedom for one main effect times the degrees of freedom for the other main effect. For the example, this is $df_{interaction} = 1 \times 1 = 1$. The total degrees of freedom is $n_t - 1$, which, for the example, is $32 - 1 = 31$. The error degrees of freedom is the total degrees of freedom minus the degrees of freedom for each mean effect and interaction, which, for the example, is $31 - 1 - 1 - 1 = 28$. The sums of squares (generated by *Minitab*), degrees of freedom, and variances (mean squares, MS) are arranged in an ANOVA table, as shown in table 12.9. Double check that the numbers make sense and agree with above. Also check calculations of the variance ratios (F).

Table 12.9   Two-Way Factorial ANOVA Using *Minitab* for Example 11.3. Raw data are shown in tables 12.7 and 12.8.

ANOVA:WeightGain versus Diet, Stress

```
Factor      Type     Levels      Values
Diet        fixed         2      C, J
Stress      fixed         2      H, L

Analysis of Variance for WeightGain

Source      DF       SS       MS       F       P
Diet         1   1568.0   1568.0   32.45   0.000
Stress       1   1458.0   1458.0   30.17   0.000
Diet*Stress  1    312.5    312.5    6.47   0.017
Error       28   1353.0     48.3
Total       31   4691.5

S = 6.95136 R-Sq = 71.16% R-Sq(adj) = 68.07%
```

As before, each MS is determined by dividing the SS by the associated df. The F statistics are determined by dividing each MS by the MS_{error}. For example:

$$F_{stress} = \frac{MS_{stress}}{MS_{error}} = \frac{1458}{48.32} = 30.18$$

We compare the calculated F ratios with the critical F values from table A.6. Since 28 df is not tabulated, we find the critical values from the table which bracket 1 and 28 df. These would be $F_{1, 25} = 4.24$ and $F_{1, 30} = 4.17$.

Since all three F statistics have the same number of degrees of freedom associated with them (table 12.9), the critical value is the same. Inspection of the table indicates that all three F statistics exceed the critical value of 4.24, so we have grounds to reject all three null hypotheses. Using table A.6, we can only find critical values at $\alpha = 0.05$, so using this table alone would only give us p-values of $p < 0.05$. However, notice that the F statistics for the two main effects are much larger than the critical value, suggesting that their p-values are much less than 0.05. Indeed, *Minitab* reports $p = 0.000$ (which actually is a small number—not zero! $p < 0.001$) for each main effect (diet, stress). We thus can conclude that there are significant effects of both diet and stress on weight gain in these mice.

Finally, we may also reject the last H_0 (no interaction) and conclude that there is a significant interaction of diet and stress on weight gain. The highest weight gain was induced in individuals exposed to the junk food diet and high stress, and the lowest weight gain was induced in individuals exposed to the regular diet and low stress. Results of this nature are sometimes easier to interpret graphically, as shown in figure 12.1.

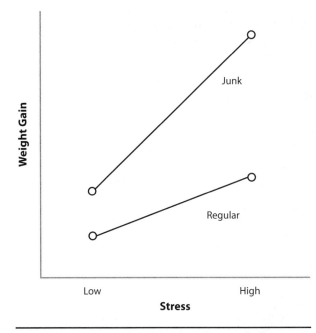

Figure 12.1 An interaction plot. The graph illustrates the cell means for example 12.4.

This figure is an example of an **interaction plot**, which illustrates the means of the response variable for each treatment combination. Connecting the means from different levels of factor 1 (in this case, stress) allows us to see if the response (change in mean with change in level) is similar for different levels of factor 2 (diet). No interaction would be indicated when the lines drawn between the means are nearly parallel. In this case, the lines are not parallel. Indeed, the weight gain to junk food appears to be enhanced in a high stress environment. Furthermore, the data show that these imaginary mice showed smaller weight gains in response to stress when fed the regular diet than fed the junk food diet. In general practice, an interaction plot is very useful to show when the interaction effect is significant.

If you have computer software available, try running the ANOVA on this example and check that your ANOVA table comes out with similar F statistics. Also, produce an interaction plot. Finally, check the assumptions of ANOVA (section 11.4). You should see that both the normality and homogeneous variance assumptions were met.

The factorial design is not restricted to two levels of each treatment, as in the preceding example, but may include as many levels of the treatments as desired by the investigator. Example 12.5 illustrates a factorial design using two levels of one main effect and three levels of a second main effect. This could be called a 2×3 factorial design.

Example 12.5
A Factorial Design with More Than Two Levels of One of the Main Effects

An evolutionary biologist selected samples of road warblers (museum specimens) of both sexes from three locations: Eastern North America, Western North America, and the Intermountain region of North America. Ten individuals were selected at random from each combination of sex and region, and their culmen (bill) lengths were measured. The results were as follows.

	Eastern	Western	Intermountain
Females	50.1	53.4	54.0
	52.8	55.2	49.1
	50.8	51.0	60.5
	58.8	59.3	57.8
	59.7	61.5	48.7
	49.0	61.2	57.0
	58.8	57.8	61.1
	62.2	50.1	62.8
	57.8	56.0	59.8
	61.2	56.5	60.3
Males	46.5	57.5	49.1
	44.4	59.3	51.8
	42.0	62.4	55.3
	51.1	61.1	43.6
	45.8	59.9	50.1
	46.3	55.6	51.0
	41.8	56.8	49.0
	52.0	59.2	48.8
	46.5	50.4	52.0
	39.0	47.8	43.0

The *Minitab* solution to this problem is given in table 12.10 on the following page. You should be able to interpret the results. Also, draw an interaction plot (like fig. 12.1), plotting location on the *x* axis and different lines for the two genders. Interpret.

Table 12.10   Two-Way Factorial ANOVA for the Bill Length Data (example 12.5)

ANOVA: BillLength versus Location, Gender

```
Factor        Type    Levels     Values
Location      fixed        3     E, I, W
Gender        fixed        2       F, M

Analysis of Variance for BillLength

Source           DF          SS         MS        F        P
Location          2      335.94     167.97     8.67    0.001
Gender            1      511.58     511.58    26.41    0.000
Location*Gender   2      350.84     175.42     9.06    0.000
Error            54     1046.03      19.37
Total            59     2244.39

S = 4.40124     R-Sq = 53.39%     R-Sq(adj) = 49.08%
```

12.4 Other ANOVA Designs

Analysis of Variance (ANOVA) is a highly versatile statistical procedure. Since its invention by R. A. Fisher, ANOVA has been widely applied in agricultural experiments and research in ecology. We have just barely scratched the surface on uses of this technique.

In chapter 11, we explored the layout and applications of one-way ANOVA and a related nonparametric procedure. In the current chapter, we explored the modification of ANOVA for related samples (the randomized block design) and the extension for analyzing the simultaneous effects of two factors on the response variable (two-way factorial ANOVA). The advantages of the two-way design are efficiency (run two experiments at once) and also the ability to check for interaction effects.

ANOVA can be applied to many other research designs. Factorial ANOVA designs have been extended to three and even four factors manipulated simultaneously (e.g., Box et al. 1978). Although informative, these experiments can be quite complicated to run and, even worse, difficult to interpret because of the many possible interaction effects. Interactions may cloud our ability to detect and interpret main effects. Additional experimental designs using ANOVA include the split plot, the Latin square, and nested ANOVA.

ANOVA can also be combined with regression analysis to compare two or more regression lines, a procedure called Analysis of Covariance (discussed in section 14.10).

Students who expect to use ANOVA in their research are encouraged to read further in one of the advanced statistical texts, such as are listed in the references (appendix C). In particular, Box et al. (1978), Snedecor and Cochran (1980), Quinn and Keough (2002), and Sokal and Rohlf (2012) have detailed treatments of ANOVA.

Key Terms

> factorial design
> interaction effect
> interaction plot
> randomized block design
> repeated-measures ANOVA

Exercises

For the following problems, unless instructed otherwise, assume that the residuals are normally distributed and variances homogeneous. For each problem, recognize the ANOVA design involved, state the null hypothesis or hypotheses, and, using statistical software, construct an ANOVA table. State a biological conclusion based on your statistical analysis. Where appropriate, construct an interaction plot and interpret. All exercises are suitable for a computer solution.

Exercises 12.1 through 12.6 have a randomized block design.

12.1 Three mice from each of six litters were randomly assigned to three treatment groups. One group was exposed to 100 ppb methyl mercury, one group was exposed to 100 ppb mercuric chloride, and the third group served as a control. The time (in minutes) that each mouse spent on an exercise wheel in one day was measured several days after exposure. Does exposure to mercury affect the mice activity levels? Do mice show variation among litters in activity levels?

Litter Number	Methyl Mercury	Mercuric Chloride	Control
1	60	90	50
2	100	120	80
3	40	45	35
4	120	110	100
5	80	105	60
6	130	155	105

12.2 An investigator wished to test the effect of 3 diets on the peristaltic blood pressure of mangrove toads. Because these animals are rare, only a few may be collected in any one location. 3 toads were collected from each of 5 locations and one toad from each location was randomly assigned to one of the 3 diets. Is there a difference in blood pressure that is related to the three diets?

Location	Diet A	Diet B	Diet C
1	65	75	80
2	60	69	79
3	55	50	70
4	53	54	80
5	64	69	85

12.3 Photosynthesis in 5 tobacco plants was measured before, one day after, and one week after exposure at a temperature of 100°F for two hours. We wish to know if exposure to this temperature affects the rate of photosynthesis.

Photosynthesis (μM CO_2/min/g)

Plant Number	Before Treatment	One Day After Treatment	One Week After Treatment
1	127	107	130
2	130	111	127
3	240	222	250
4	116	98	120
5	215	201	200

12.4 A parcel of land was divided into 6 equal-sized plots (blocks), and each block was divided into 3 equal-sized subplots. 3 treatments (no added nitrogen, 10 lbs nitrogen/hectare, and 100 lbs nitrogen/hectare) were randomly assigned to the subplots within each plot. Does added nitrogen increase the growth of this crop? Data are pounds of yield per subplot.

	Crop Yield With		
Block	No added N	10 lbs/ hectare	100 lbs/ hectare
1	105	156	187
2	98	145	167
3	125	170	201
4	100	150	180
5	130	185	210
6	80	135	162

12.5 The concentration of unicellular algae (measured as chlorophyll concentration in $\mu g/L$) at three different depths in four lakes was measured. We wish to know if there is a difference in algae concentration that is related to depth. Lakes are treated as blocks to take into account differences among lakes.

Lake	Surface	1 m	3 m
1	425	130	56
2	500	215	115
3	100	30	10
4	325	100	28

12.6 A plant ecologist wished to determine if an exotic weed might be becoming more numerous in a particular area. She randomly selected 5 1m² quadrats in the area of interest and counted the number of individual plants of the weed in each quadrat over a period of 5 years. Is there a significant change in density of this species over this time period? By inspecting the data, does there appear to be an increase over time?

Quadrat Number	Year 1	2	3	4	5
1	2	5	8	10	20
2	5	9	17	30	40
3	15	30	31	60	72
4	0	2	9	15	24
5	9	11	23	17	45

Exercises 12.7 through 12.8 are appropriate for the Friedman two-way ANOVA.

12.7 Six male mosquito fish were selected at random from a large population. Each was placed individually in an aquarium with a female, and the number of copulatory attempts made during 5 successive three-minute intervals was recorded. The results were as follows. We wish to know if males exhibit a diminishing response (habituate) to individual females. These data do not have homogeneous variances and are unlikely to be fixed by transformation.

Male Number	Interval Number				
	1	2	3	4	5
1	20	10	7	5	2
2	18	7	5	5	3
3	25	13	9	4	1
4	10	5	3	2	1
5	12	15	8	3	0
6	17	4	5	2	1

12.8 Three randomly selected mice from each of 6 litters were randomly assigned to 3 treatment groups. One group received injections of 100 μg/kg of epinephrine, the second group received injections of 100 μg/kg of norepinephrine, and the third group served as a control. SRBC antibody titers were measured one week after treatment. The variable does not appear to be normally distributed, nor are the error variances equal. Does the type of treatment affect the antibody titer level?

Litter Number	Epinephrine	Norepinephrine	Control
1	125	180	350
2	225	112	400
3	180	290	375
4	300	225	495
5	115	325	427
6	98	115	510

Exercises 12.9 through 12.12 have a two-way factorial design.

12.9 The possible influence of crowding and sex on plasma corticosterone in a highly inbred strain of rug rats was investigated using a factorial design. Sex (males, nongravid females, and gravid females) and crowding (low, moderate, and high) were used as the main treatment effects. The results were as follows. Test whether there is an effect of sex, an effect of crowding, and an interaction effect of these two factors on plasma corticosterone levels (measured in ng/ml).

Sex	Crowding		
	Low	Moderate	High
Males	5	115	253
	8	122	249
	13	119	260
	9	130	257
	15	114	280
	11	129	263
Nongravid Females	12	112	219
	19	115	222
	15	121	218
	20	117	220
	11	118	223
	18	120	225
Gravid Females	37	157	289
	42	160	273
	50	173	280
	35	182	291
	40	168	205
	36	170	296

12.10 Forty-five large female guppies were randomly assigned to 9 groups, each of which received different amounts of food and were kept at different temperatures in a two-way factorial design. The number of offspring that each produced in a single brood are given in the following table. We wish to know if there is an effect of temperature, an effect of food intake, and an interaction of these two factors on the number of offspring per brood produced by these animals.

Temperature	Number of Daily Feedings		
	1	2	3
70°F	18	25	28
	20	30	36
	15	19	29
	27	30	30
	30	25	37
75°F	20	28	33
	28	29	39
	30	32	42
	17	38	47
	29	29	38
80°F	35	35	51
	30	39	42
	32	30	48
	28	40	39
	35	38	55

12.11 The possible effects of sex and age on systolic blood pressure (mm Hg) in hamsters were investigated, with the following results. Determine if there is an effect of sex, an effect of age, and an interaction of these two factors.

| Age | Sex | |
---	Male	Female
Adolescent	108	110
	110	105
	90	100
	80	90
	100	102
Mature	120	110
	125	105
	130	115
	120	100
	130	120
Old	145	130
	150	125
	130	135
	155	130
	140	120

12.12 Using the data from digital table 3, determine if there is an effect of smoking, an effect of sex, and an interaction between these two factors on pulse rate in humans.

Choose the appropriate ANOVA design for exercises 12.13–12.17

12.13 Six rug rats were given a small amount of caffeine. Their pulse rate (beats per minute) was measured before, immediately after, and one hour after administration of the caffeine. Does caffeine affect pulse rate in this species?

Rat Number	Before	Pulse Rate Immediately After	One Hour After
1	105	115	108
2	98	110	100
3	110	125	115
4	100	112	105
5	114	130	120
6	90	100	95

12.14 The effect of infection by tobacco mosaic virus (TMV) and tobacco ringspot virus (TRSV) on o-diphenol oxidase activity in three genetic strains of tobacco was measured. The results are given in the following data table.

| | Treatment | | |
	Non-infected	TMV-infected	TRSV-infected
Strain A	102	237	117
	115	219	95
	98	201	128
Strain B	97	193	105
	85	175	60
	63	160	91
Strain C	127	230	135
	150	249	145
	168	250	170

Answer the following questions and support your answers with the appropriate ANOVA terms:

1. Does virus infection affect this enzyme?
2. Is there a difference in the activity of this enzyme among strains?
3. Do different strains respond differently to virus infection (i.e., is there a significant interaction between the two main effects)?

Show the interaction results graphically.

12.15 Six damselfly larvae were placed individually into 6 containers, which contained 10 each of 4 different prey species. After a short time, the number of each prey species eaten was recorded. Is there a preference for any of the prey items? Note that we expect these data not to have homogeneous variances.

Naiad Number	Prey A	Prey B	Prey C	Prey D
1	8	3	1	2
2	9	5	4	1
3	7	1	0	0
4	10	7	8	1
5	8	9	1	2
6	9	2	1	3

12.16 Using the data from digital table 3, determine if reaction time is different between males and females, between athletes and nonathletes, and if there is an interaction between gender and athletic participation.

12.17 Testosterone levels in 6 captive male bush hogs was measured at 4 times during the year. The mating season for this species occurs in the fall. Do the data below support the hypothesis that testosterone levels in this species are higher during the mating season?

Plasma Testosterone Level

Male Number	Winter	Spring	Summer	Fall
1	20	30	25	220
2	30	20	70	210
3	40	90	50	230
4	35	40	40	190
5	60	60	105	100
6	55	40	72	210

Associations between Two Measurement Variables
Correlation

Often in biological research we are interested in exploring the possible relationships between two or more measurement variables. Examples include the relationship between human blood pressure and the risk of stroke, the size of female lizards and the number of offspring they produce, and the relationship between drug dosage and a particular physiological response. In all of these cases, two variables are measured for each individual in the sample, and we seek to detect a relationship between these two variables. Notice the distinction from one-way ANOVA (chapter 11), where one of the variables (the treatment factor) is categorical and the other (the response) is a measurement variable.

With two measurement variables, we can analyze the data in an assortment of ways. Before doing anything else, *always plot your data*. Sometimes this is all we need do. For instance, we might simply want to explore some trends over space or time (figs. 13.1 and 13.2 on the following page).

In other situations, we use statistical methods to test hypotheses or model the relationship between the two variables (fig. 13.3). Two widely used methods are correlation and regression, which are the subjects of the following two chapters. The particular technique that should be used depends on the particular questions being asked about the data.

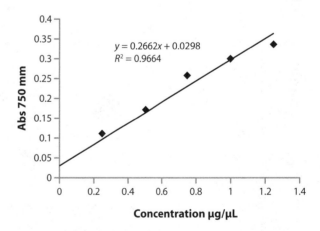

Figure 13.3   Example of a regression analysis: standard curve for protein concentration.
Abs = −0.0846 + 3.64 Conc
R^2 = 96.1%

129

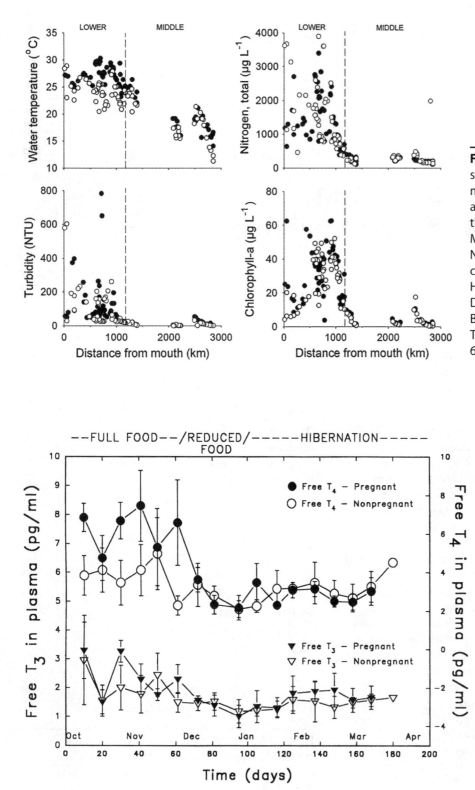

Figure 13.1 Examples of scatterplots for several variables measured on the Missouri River, as a function of distance from the confluence with the Mississippi at St. Louis (km 0). Notice that the dots are not connected with lines. Source: Havel, J. E., K. A. Medley, K. R. Dickerson, T. R. Angradi, D. W. Bolgrien, P. A. Bukaveckas, and T. M. Jicha. 2009. *Hydrobiologia* 628: 121–135.

Figure 13.2 Example of scatterplots illustrating trends with time for two variables (thyroxin fractions T3 and T4) classified by group (pregnant or nonpregnant female bears). Notice that for time series such as this, the dots are connected with lines. The bars above and below the means are the standard errors of the means (SE). Source: Tomasi, T. E., T. J. Tucker, and E. C. Hellgren. 1998. *General & Comparative Endocrinology* 109:192–199.

13.1 Associations or Modeling: Correlation and Regression

Correlation analysis is used only to determine the degree of association between two measurement variables. In contrast, **regression analysis** is used to model the dependence of one variable (the response variable) on the other variable (the predictor). With regression we assume a cause-and-effect relationship between the two variables, such that a substantial proportion of the variation in the response variable can be explained by the predictor variable.

For example, we might investigate a possible dependence of pulse rate on caffeine consumption. If such a relationship exists, it is the amount of caffeine consumed that causes a change in pulse rate, and not the other way around! Quite commonly, the independent variable (predictor) is not a random variable but is rather under the control of the investigator. The experimenter chooses what specific concentrations of caffeine to administer to the test subjects. Because of its widespread use by biologists, regression analysis will be described in some detail in chapter 14.

In **correlation analysis** we are only interested in knowing the strength of association between two measurement variables. We do not attempt to model one variable on the other. With correlation, we ask two questions: (1) Are two measurement variables related in some consistent and linear way and, if so, in what direction? and (2) What is the strength of the relationship? The strength of such relationships is reflected by how closely the points in a scatterplot of the two variables cluster about an imaginary line drawn through them. (Note that we do not actually draw this line, and it would be improper to do so in a correlation analysis.) With correlation, there is no assumption about a "cause-and-effect" association between the two variables, although such a relationship might exist. For example, we might wish to know if student scores on exams are associated with those of their take-home problem sets. In other words, do students who do well on problem sets also tend to do well on exams? We are not attempting to attribute causation of one variable from another. Both may instead be responding to other variables, such as class attendance or time spent working practice exercises. We only want to know how strongly they are associated. Other examples of correlation include: Do two methods of measuring blood pressure tend to give similar results? How strongly associated are pairs of morphometric characteristics of grizzly bears? Is there a correspondence between concentrations of two toxic metals (e.g.,

cadmium and lead) in the sediments of streams in a watershed impacted by industrial pollution? These questions all ask for correlation analysis. In an earlier chapter, we explored a similar question about the association between two variables, except using categorical variables with frequency data (section 7.2).

13.2 The Pearson Correlation Coefficient

The measure of the strength of the relationship between two variables in a correlation is the **correlation coefficient**, formally called the Pearson correlation coefficient (or also the product moment correlation coefficient). The parameter of the population is designated by the Greek symbol rho (ρ). Usually, the true value of this parameter is unknown to us, and we must estimate its value from a random sample of the population. The sample correlation coefficient is designated as r. The value of ρ (and r) ranges from +1, indicating a perfect positive correlation (all points falling one line with a positive slope); through 0, indicating no relationship between the two variables; to –1, indicating a perfect negative correlation between the two variables. See figure 13.4 on the next page which illustrates different degrees of association between pairs of variables. Notice that we are not saying anything about the value of the slope. In fact, with correlation, slope is entirely irrelevant to the question. We only want to answer questions 1 and 2 above.

The null hypothesis in a correlation analysis is that the parametric correlation coefficient, ρ, is zero, or

$$H_0: \rho = 0$$

When we are able to reject H_0, we conclude that ρ is not equal to zero and that therefore a correlation or association between the two variables exists.

Assumptions of the Test
1. The sample is a random sample from the population of interest.
2. Each variable is a continuous measurement that follows a normal distribution.
3. The relationship between the two variables, if it exists, is linear.

When assumptions 2 and/or 3 are not met, transformation of one or both variables may correct the data (see section 11.5.1). If not, a nonparametric

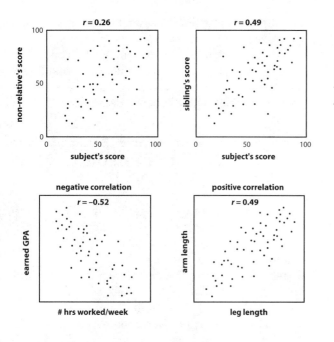

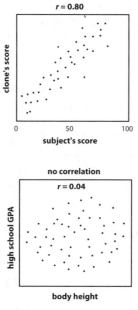

Figure 13.4 Scatterplots and correlation statistics for six different datasets showing different degrees of association between two variables. Adapted from J. Sumich, unpublished.

correlation test should be used instead (section 13.4 below). When assumption 1 is not met, correlation analysis is inappropriate. The data can't be "fixed," and so a new sample must be collected.

To illustrate the calculations involved in correlation analysis, we use a hypothetical example of heavy metal concentrations in stream sediment samples. These data are shown below in table 13.1 with some preliminary calculations and are also illustrated in figure 13.5. First consider this example. We are not interested in knowing about causation; we only want to know if there is a tendency for one variable to increase (or to decrease) as the other variable increases.

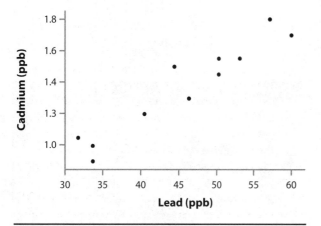

Figure 13.5 Association between cadmium and lead concentrations in hypothetical stream sediment samples. (Example 13.1)

Table 13.1 Heavy Metal Concentrations in Hypothetical Stream Sediments

Sample number	Cadmium (ppb)	Lead (ppb)
1	0.90	33
2	1.55	50
3	1.30	46
4	1.00	33
5	1.55	53
6	1.80	57
7	1.50	44
8	1.05	31
9	1.70	60
10	1.20	40
11	1.45	50

$\Sigma x = 15$ $\Sigma y = 497$
$\Sigma x^2 = 21.33$ $\Sigma y^2 = 23,449$
$(\Sigma x)^2 = 225$ $(\Sigma y)^2 = 247,009$
$\Sigma xy = 705.8$
$\Sigma x \Sigma y = 7,455$

In this example, the data are continuous measurements, and both variables may be assumed to be approximately normally distributed. Accordingly, we may compute the **Pearson correlation coefficient** (r) by

$$r = \frac{\Sigma xy - \frac{\Sigma x \Sigma y}{n}}{\sqrt{\left(\Sigma x^2 - \frac{(\Sigma x)^2}{n}\right)\left(\Sigma y^2 - \frac{(\Sigma y)^2}{n}\right)}} \quad (13.1)$$

The numerator in equation 13.1 is called the covariance, which measures how x and y vary together. The denominator is the square root of the product of the sums of squares for each of the variables. Substituting the values from the example in equation 13.1 gives

$$r = \frac{705.8 - \left[\dfrac{(15)(497)}{11}\right]}{\sqrt{\left(21.33 - \dfrac{(15)^2}{11}\right)\left(23,449 - \dfrac{(497)^2}{11}\right)}} = 0.952$$

Recall that the correlation coefficient can range from –1 to +1. The value 0.952 is close to 1.00, suggesting a very strong positive correlation between the two variables. (As the concentration of one metal increases, the concentration of the other metal also increases.) Referral to figure 13.5 above indicates this is indeed the case. Nevertheless, we should always compare the sample statistic with critical values to assess the original null hypothesis.

13.2.1 Testing the Significance of r

Recall that the null hypothesis in a correlation analysis is that ρ, the population correlation coefficient, is zero. The sample correlation coefficient, r, is an estimate of this population correlation coefficient. When the null hypothesis is that the population correlation is zero (the usual case), r has a t distribution with $n - 2$ degrees of freedom. We calculate t by

$$t = r\sqrt{\frac{n-2}{1-r^2}} \tag{13.2}$$

For the example,

$$t = 0.952\sqrt{\frac{9}{1-0.906}} = 9.315$$

Referring to table A.2, the critical value of t for alpha = 0.05 and 9 df is 2.262 for a two-tailed probability. Since our t statistic exceeds even the largest critical value at 9 df, the p-value is very small ($p < 0.0001$). We may therefore reject the null hypothesis and conclude that concentrations of the two metals are correlated.

An alternative way of determining if r is significant is to consult table A.8, "Critical Values of the Pearson Correlation Coefficient (r)." This table gives the minimum values of r that permit one to reject the null hypothesis. If the calculated value of r is equal to or greater than the critical value for the specified degrees of freedom ($n - 2$), the null hypothesis is rejected. In the current example, 0.952 > 0.602, leading to the same conclusion as above: concentrations of cadmium and lead show a strong positive correlation in these stream sediments.

13.3 A Correlation Matrix

Quite often we have multiple measurement variables and would like to quickly assess which are associated with which. We might have measures of a number of environmental features and would like to know how each feature tends to associate with all other features. If we had 10 such features, there are 45 possible pair-wise combinations. This leads to a lot of number crunching. Fortunately, computers are good at this.

Computer statistical packages can do correlations quite easily and typically show the results in the form of a matrix, with the statistic (r) and p-value illustrated for each pair. Such a result is shown in table 13.2 on the following page for student scores from a recent biometry class ($n = 55$). The five variables are listed at the top, together with their descriptive statistics. The correlation matrix is shown at the bottom for all possible pairs of these variables. (Note that one of the variables is derived from summing three of the others.) In this matrix, at each cell (an intersection of row and column position) the top number is the correlation coefficient and bottom number the p-value. The p-values all being 0.000 (actually $p < 0.001$) indicate that every variable is strongly correlated to every other one. Furthermore, the positive values for the correlations indicate that as one score increases other scores tend to increase. A scatterplot for two of these variables is shown in figure 13.6 on the next page. Notice that, although the positive relationship between these two variables is evident, substantial unexplained variation remains.

One point of caution about correlation matrices is that the risk of a type I error increases with multiple tests (recall section 11.3.2). The simplest solution is to divide alpha (usually 0.05) by the number of tests. In this case, there are 10 pairwise combinations of five variables, so we would reject H_0 ($\rho = 0$) only when the p-value is less than $0.05 \div 10 = 0.005$. That is clearly the case for all our comparisons (table 13.2), so our earlier conclusions still hold.

Table 13.2 Descriptive Statistics and Correlation Matrix for Scores by 55 Biometry Students, as Displayed by *Minitab.* **For each pair of variables in the correlation matrix, the top value is Pearson's correlation coefficient (r) and the bottom value is the p-value. A p-value reported as 0.000 should be interpreted as $p < 0.001$. For $n = 55$, critical values of r: $r_{53, .05} = 0.264$, $r_{53, .01} = 0.365$.**

Descriptive Statistics: Exam1, Exam2, Exam3, ProbSetTotal, ExamsTotal

Variable	N	N*	Mean	SE Mean	StDev	Minimum	Q1	Median	Q3
Exam1	55	1	81.55	1.41	10.42	59.00	74.00	86.00	89.00
Exam2	55	1	81.45	1.56	11.58	58.00	74.00	83.00	91.00
Exam3	54	2	79.61	1.49	10.95	47.00	73.75	82.00	88.00
ProbSetTotal	55	1	103.05	2.23	16.57	53.00	94.00	109.00	115.00
ExamsTotal	54	2	243.33	3.85	28.32	172.00	223.50	245.00	269.25

Variable	Maximum
Exam1	95.00
Exam2	100.00
Exam3	98.00
ProbSetTotal	120.00
ExamsTotal	282.00

Correlations: Exam1, Exam2, Exam3, ProbSetTotal, ExamsTotal

	Exam1	Exam2	Exam3	ProbSetTotal
Exam2	0.654			
	0.000			
Exam3	0.597	0.685		
	0.000	0.000		
ProbSetTotal	0.494	0.521	0.493	
	0.000	0.000	0.000	
ExamsTotal	0.844	0.893	0.876	0.541
	0.000	0.000	0.000	0.000

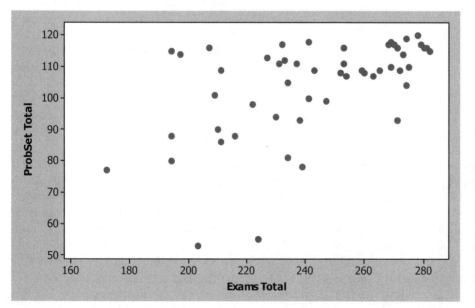

Figure 13.6 *Minitab*-generated scatterplot illustrating correlation of scores from problem sets against scores from exams for 55 biometry students. $r = 0.541$, $p < 0.001$. This is one pair chosen from the correlation matrix shown in table 13.2.

13.4 Nonparametric Correlation Analysis (Spearman's *r*)

When data are measured on an ordinal scale, or when other assumptions of the parametric correlation test are not met, one may use one of several nonparametric correlation tests. Most commonly used is the Spearman rank correlation test. The assumptions of this test are that observations are a random sample from the population and that measurement of both variables is at least ordinal.

Example 13.2
A Nonparametric Correlation

The mass (in grams) of 13 adult male tuatara and the size of their territories (in square meters) was measured. Are territory size and the size of the male holding the territory related? (In other words, do larger males hold larger territories?)

Although mass of these animals is normally distributed, the area of their territories is probably not. Furthermore, territory size is rather difficult to measure accurately. For these reasons a nonparametric correlation test seems to be in order. The results are shown in table 13.3. Here we arbitrarily designate mass as the *x* variable and territory size as the *y* variable. A scatterplot (not shown) suggests a positive association between these variables.

Table 13.3 Mass and Territory Size of Adult Male Tuataras

Observation Number	Mass (x)	Rx	Territory Size (y)	Ry	d	d²
1	510	6	6.9	6	0	0
2	773	9	20.6	11	−2	4
3	840	13	17.2	9	4	16
4	505	5	6.7	5	0	0
5	765	8	20.0	12	−4	16
6	780	10	24.1	13	−3	9
7	235	1	1.5	1	0	0
8	790	11	13.8	8	3	9
9	440	3	1.7	2	1	1
10	435	2	2.1	3	−1	1
11	815	12	20.2	10	2	4
12	460	4	3.0	4	0	0
13	697	7	10.3	7	0	0
						$\Sigma d^2 = 60$

Data from J. Gillingham

The steps involved in calculating the Spearman correlation coefficient r_s follow. We will use the tuatara data (table 13.3) as an example.

1. Rank the observations in the *x* variable from smallest to largest. These are designated as R*x* in the table.

2. Rank the observations in the *y* variable from smallest to largest. These are designated as R*y* in the table. (Note: The *x* and *y* variables are ranked separately.)

3. Subtract each R*y* from its corresponding R*x* (designated as *d* in the table), and square each difference (designated as d^2).

4. Sum the values of d^2, which is designated as Σd^2. For the example, this value is 60.

5. Calculate r_s using equation 13.3.

$$r_s = 1 - \left(\frac{6 * \Sigma d^2}{n * (n^2 - 1)} \right) \qquad (13.3)$$

where *n* is the number of observations.

For the example,

$$r_s = 1 - \left(\frac{6 * 60}{13 * (169 - 1)} \right) = 0.835$$

The null hypothesis in the Spearman correlation test is essentially the same as it is for the Pearson (parametric) test, which is that there is no relationship between the two variables. We may test the significance of r_s by using equation 13.2 in the same manner that the significance of the Pearson correlation coefficient was tested. For this example,

$$t = 0.835 \sqrt{\frac{13 - 2}{1 - (0.835)^2}} = 4.89$$

If our computed value of *t* is equal to or greater than the critical value of *t* for the desired level of alpha at $n - 2$ degrees of freedom (see table A.2), we may reject H_0.

The null hypothesis may also be tested by consulting table A.9. This table gives minimum values of r_s, which are significant at various degrees of freedom. When the calculated value of r_s is equal to or greater than the table value, the null hypothesis is rejected. Using either approach, you should see that the null hypothesis is rejected. Using table A.2 (at 11 df), we see that $p < 0.001$. Clearly, there is a tendency for larger tuatara males to hold larger territories. According to behavioral ecologists, such a result is consistent with studies from a wide variety of other animals.

Key Terms

correlation coefficient
correlation matrix
Pearson correlation
regression analysis
Spearman rank correlation

Exercises

For exercises 13.1 through 13.6, compute the Pearson correlation coefficient, test the null hypothesis $\rho = 0$, and draw the scatterplot of the data. Assume a bivariate normal distribution. Exercises may be done either by hand or with statistical software.

13.1 The systolic blood pressure and diastolic blood pressure for a random sample of 62 people were determined. Is there a correlation of systolic blood pressure and diastolic blood pressure?

Systolic	Diastolic	Systolic	Diastolic
114	74	112	68
118	68	111	70
94	54	90	53
94	48	92	56
118	64	120	70
140	70	120	80
118	78	138	84
120	80	123	78
100	40	132	83
125	90	110	77
108	58	102	66
130	90	108	74
130	94	140	86
104	62	132	72
134	72	138	82
122	86	122	78
142	82	132	64
122	62	125	63
110	72	90	42
110	64	122	78
110	68	106	62
138	82	118	78
118	62	120	78
118	76	132	76
114	78	118	80
112	78	115	70
116	79	114	70
117	78	106	66
114	76	114	66
116	74	110	70
102	62	80	42

13.2 It is likely that students who do well on an examination in a particular course are likely to score well on a subsequent examination in that same course, and that students who do not do well on the first exam are likely to not do well on the second exam (i.e., exam scores on the two exams will be correlated). Below are the scores on two examinations for 45 randomly selected students in a general biology course. Is there an association between the scores on the two tests?

Test 1	Test 2	Test 1	Test 2	Test 1	Test 2
63	76	49	71	66	67
51	48	46	69	72	57
46	64	54	62	74	74
74	88	74	88	74	83
83	88	40	64	83	83
80	86	74	83	83	86
49	79	72	60	83	83
89	79	72	62	80	57
77	74	51	52	80	62
86	79	66	74	72	50
46	79	89	88	83	69
60	74	60	79	77	60
66	83	74	64	92	88
66	71	72	74	86	74
54	81	69	76	60	62

13.3 Ten randomly selected soil samples were analyzed for krypton content (µg/kg soil) using an old, expensive, but very reliable and accurate method; and a newer, less expensive and faster method. A strong correlation between the results of the two methods would indicate that the new method is also accurate.

Old Method	New Method
25	27
30	28
20	19
35	36
40	38
25	25
33	32
50	52
65	67
60	58

13.4 Using the data in Digital Appendix 3, determine if there is a correlation between pulse rate and reaction time in humans.

13.5 Total body weight, spleen weight, and bursa weight of nine newly hatched turkeys were determined. Is there a correlation between total body weight and spleen weight? Total body weight and bursa weight? Spleen weight and bursa weight?

Body Weight (in grams)	Spleen Weight (in milligrams)	Bursa Weight (in milligrams)
53.81	18.9	50.8
56.26	20.4	51.4
59.86	15.9	28.4
59.96	19.9	66.6
61.75	17.4	35.5
55.28	24.0	38.8
56.57	21.3	50.3
49.91	16.2	33.2
54.25	19.3	39.3

13.6 The weight, length, and width of 44 randomly selected killdeer eggs were determined. Is there a correlation between weight and width? Width and length? Weight and length?

Weight (g)	Width (mm)	Length (mm)	Weight (g)	Width (mm)	Length (mm)
13	26.4	36.7	13.5	27.4	37.4
13	26.5	36.5	13.5	26.9	38.5
12	26.4	34.3	14	27.4	37.8
13.5	27.1	37.1	15.5	28.7	38
15.5	28.3	38.1	15.5	28.4	39
15	28	37.2	16	28.4	39
15	28	38.1	12.5	26.3	37.6
14	27.2	37.5	13	26.5	37.6
14	27.7	36.7	12	25.9	37.1
15	27.6	38.3	14	27.1	38.3
14.5	28	36.5	15	27	42
14.5	27.2	39.5	14.5	27.3	39.6
14.5	26.8	39.1	14	27	37.8
13	26	39.6	15	27.2	40.2
13	26	36.9	12	27.5	35.3
12.5	27.3	36.4	12.5	27.3	34.3
12	26.7	36	13	26.3	38.1
12.5	27.3	36.5	13	26.4	38.2
12.5	26.8	37.1	13	28.3	39.3
14.5	27.2	37.6	14	29.4	38.7
14.5	27.4	39.5	15	28.4	39.6
13.5	27.8	38.1	14	27.7	39.3

Data from D. Blaszkiewicz

For exercises 13.7–13.8, you may not assume normal distributions. Prior studies have shown that such behavioral data tend to show strong departures from normality. Inspect a scatterplot to describe trends. Then use the appropriate nonparametric test to evaluate the null hypothesis of interest.

13.7 A study tested the hypothesis that larval ringed salamanders respond to chemical cues from predatory newts and that smaller salamanders have a greater response to the cues than larger salamanders. The "change in activity" of salamanders in response to the odor of

predatory newts was measured, with positive numbers indicating increased activity and negative numbers indicating decreased activity. Do smaller salamanders tend to show the largest decrease in activity when exposed to the odor of predatory newts? Is there a significant association between the two variables?

Size (mm)	Change in activity	Size (mm)	Change in activity
2.4	65	2.3	−148
1.5	−7	2.2	−12
2.7	99	2.5	−69
2.5	75	2.6	−32
2.4	−55	2.0	18
2.5	121	2.8	68
2.0	−47	2.0	−188
1.9	−150	2.1	−248

Data from A. Mathis

13.8 In black swans, both sexes possess curled feathers on their wings. A study on the adaptive significance of this trait investigated whether this trait is important in social pair formation. Based on the following data, is the number of curled feathers correlated between members of a social pair?

Number of curled feathers	
Female	Male
26	20
22	22
26	23
23	25
26	26
20	31
28	27
27	27
31	30
27	32
27	33
25	32
27	33
29	35
32	29
36	30
37	31
38	31
38	34
38	40
33	41
35	45

Data adapted from Kraaijeveld et al. 2004

chapter fourteen

Modeling One Measurement Variable against Another
Regression Analysis

Regression Analysis is used to model the dependence of one measurement variable on another (independent) variable. In other words, we would like to develop a mathematical equation that describes the dependence of one variable on the other. The dependent variable is also called the **response variable** and the independent variable the **predictor variable**. We generally assume a cause and effect relationship whereby variation in the predictor variable causes changes in the response variable. If caffeine causes an increased heart rate then increases in coffee consumption should cause increases in heart rate. In contrast, correlation (chapter 13) is only concerned with measuring the association between variables and not developing a predictive model. In **simple linear regression** (our topic in this chapter), a single predictor variable is used to develop a linear equation that describes the relationship between the predictor and the response variables.

In regression analysis, the predictor variable is usually not a normally-distributed random variable, but is rather under the control of the investigator. In this respect regression has a **fixed-effects** design (sometimes called model I regression), which is analogous to a fixed-effects ANOVA (see section 11.3). Fixed effects means that the experimenter chooses what levels of the predictor variable to ad-

minister to the test subjects. We assume that these levels are exactly known. For the caffeine-heart rate example, caffeine concentrations are exactly known whereas heart rates are subject to some measurement error. A **random-effects** design (also called model II regression) may also be used with regression analysis. However, this model is a little trickier in practice, since both the predictor and response variables are random variables subject to random errors. Examples 14.1 and 14.2 on the following page illustrate each of these regression models.

The scatterplots for these data are shown in figures 14.1 and 14.2. Note that in figure 14.1, temperature is plotted on the horizontal axis, and heart rate is plotted on the vertical axis, and not vice versa. Furthermore, we have drawn a line through the points and could, if we choose, write an equation that describes this line. Using this equation we could even predict the heart rate of juniper pythons at various temperatures. We will explore the first example in some detail throughout this chapter. Inspection of figure 14.2 indicates that fecundity is closely related to body size (weight) and that we generated a linear equation that allows predicting fecundity from body size. This sort of analysis is commonly done by ecologists. However, we have to be careful about inferring causation from such data. Although fecundity

Example 14.1
A Fixed Effects Regression Problem

A snake physiologist wished to investigate the effect of temperature on the heart rate of juniper pythons. She selected nine specimens of approximately the same age, size, and sex and placed each animal at a preselected temperature between 2° and 18°C. After the snakes had equilibrated to their ambient temperatures, she measured their heart rates. The results are given in table 14.1. Note that, in this experiment, the temperatures were selected or under the control of the investigator. Since this is a manipulative experiment, we can safely infer that changes in temperature *cause* changes in heart rate.

Table 14.1 The Relationship between Temperature and Heart Rate in Juniper Pythons

Snake Number	Temperature (°C)	Heart Rate (BPM)
1	2	5
2	4	11
3	6	11
4	8	14
5	10	22
6	12	23
7	14	32
8	16	29
9	18	32

Example 14.2
A Random Effects Regression Problem

A herpetologist wanted to predict clutch size of iguanas based on their body size. She collected a random sample of 11 gravid female iguanas, counted the number of eggs laid (fecundity) by each lizard, and then measured the weight of each female after eggs were laid. The results are given in table 14.2. Notice that both egg number and body weight are random variables and that both variables were sampled from a survey of iguanas captured from a natural population.

Table 14.2 Postpartum Weight of Female Iguanas and Number of Eggs Produced

Specimen Number	Mass (in Kilograms)	Number of Eggs
1	0.90	33
2	1.55	50
3	1.30	46
4	1.00	33
5	1.55	53
6	1.80	57
7	1.50	44
8	1.05	31
9	1.70	60
10	1.20	40
11	1.45	50

Data from T. Miller

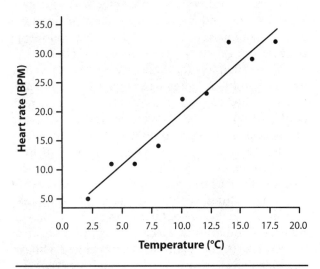

Figure 14.1 Heart rate (BPM) of juniper pythons as a function of temperature (data from example 14.1).

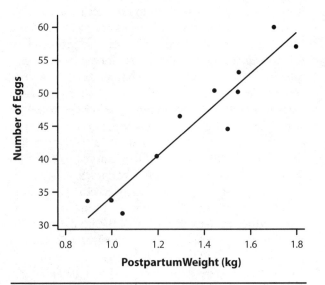

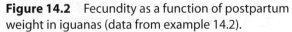

Figure 14.2 Fecundity as a function of postpartum weight in iguanas (data from example 14.2).

appears to increase with female size, we cannot say for sure that greater size necessarily caused the increase in fecundity. Both variables may be responding to another unmeasured variable, such as age or nutrition. We will not consider the details of the random effects regression procedure, although interested students can read further in books on regression (e.g., Neter et al. 1989).

14.1 Simple Linear Regression Fundamentals

In regression analysis we assume that there is a cause-and-effect relationship between the two variables under study. Furthermore, in most cases we also assume that the independent variable is under the control of the investigator (i.e., that a fixed-effects experimental design is used). In effect, we hypothesize that there is a functional relationship that permits us to predict a value of the dependent variable, y, corresponding to a given value of the independent variable, x. Mathematically, such a relationship is expressed as:

$$y = f(x)$$

In simple linear regression, the functional relationship between y and x takes the form

$$\mu_y = \alpha + \beta x \qquad (14.1)$$

where μ_y is the population mean value of y at any value of x, α is the population intercept, and β is the population slope. Notice that this equation describes a straight line. Recall other similar formulas you may have seen (e.g., $y = a + bx$ or $y = mx + b$); these are all different ways of saying the same thing—plot a straight line having a particular slope ($\Delta y / \Delta x$) and y intercept (value of y where $x = 0$). For any particular equation (say $y = 2.0 + 0.8x$), you should be comfortable showing a graph of the function (or vice versa).

Any particular value of y deviates from its expected value (μ_y) due to some unexplained variation, which we call a **residual** (e):

$$y_i = \alpha + \beta x_i + e \qquad (14.2)$$

We assume that these residuals (sometimes called error terms) have a standard normal distribution (recall section 8.2).

Regression analysis has several goals, which include but are not limited to the following.

1. Regression is used to estimate an equation that describes the linear relationship between the two variables in question. This is called the **regres-**

sion equation or the regression function. Since the parameters α and β are usually unknown to us, we estimate these values from a sample.

2. From this equation we are able to construct a line through the points of a scatterplot, which is called the **least squares regression line**.

3. The regression equation may be used to predict values of the dependent variable (y) at various values of the independent variable (x).

4. Regression may be used to estimate the extent to which the dependent variable is under the control of the independent variable. In other words, how much of the variation in y is explained by x?

Simple linear regression analysis is based on several assumptions. One of the assumptions involves which analysis method is most appropriate (assumption 1). Others deal with the nature of the data (assumptions 2, 3, and 5). Sometimes, transformations may be used to correct problems in the data (recall section 11.5).

Assumptions of the Test

1. The independent variable is fixed. This means, in effect, that values of the independent variable are chosen by the investigator and do not represent a random variable in the population. There is thus no variance associated with the independent variable. [Recall that random effects (model II) regression does not require this assumption. For more information on model II regression, see Neter et al. (1989).]

2. For any value of the independent variable (x), there exists a normally distributed population of values of the dependent variable, y. The population mean of these values of y, μ_y is

$$\mu_y = \alpha + \beta x$$

   where α is the population intercept and β is the population slope of the regression equation.

3. The variances of the residuals for all values of x are equal. (This is analogous to the homogeneous variance assumption of ANOVA.)

4. Observations are independent. In a practical sense, this means that each individual in the sample is measured only once!

5. For linear regression, we also assume that the functional relationship is linear. To check this assumption, *always plot the data.*

When assumption 1 is not met, fixed effects regression is not the proper treatment for the data. However, depending on the question, a random effects regression model might be appropriate, but requires careful interpretation (example 14.2). Alternatively, if the goal is simply to look for associations, a correlation analysis should be used instead (section 13.1).

Consider the experiment to determine the effect of temperature on heart rate in juniper pythons (example 14.1). The investigator selected a series of temperatures (assumption 1) and designated this as the independent variable. He then placed a different animal at each of the preselected temperatures and measured its heart rate (the dependent variable). The data obtained in the experiment are shown again in table 14.3, this time with some statistics added.

Table 14.3   The Effect of Temperature on the Heart Rate of Juniper Pythons

Temperature (°C) (x)	Heart Rate (BPM) (y)
2	5
4	11
6	11
8	14
10	22
12	23
14	32
16	29
18	32
$\Sigma x = 90$	$\Sigma y = 179$
$\bar{x} = 10$	$\bar{y} = 19.88$

Note that the investigator measured only one individual at each temperature and that at each temperature a different individual was used (assumption 4). Had he measured the heart rate of one individual at different temperatures, the observations would not have been independent and any inferences about a relationship between the two variables would apply only to that single individual (a trivial result).

These data are shown as a scatterplot (x-y plot) in figure 14.1 above. Note that, although not all of the points seem to fall on a straight line, there seems to be a definite linear relationship between the two variables (assumption 5). Regression analysis can tell us more about this relationship.

14.2 Estimating the Regression Function and the Regression Line

One of the things we are interested in doing with regression analysis is to estimate the paramet-

ric regression function (regression equation), $\mu_y = \alpha + \beta x$, where α is the intercept and β is the slope of the equation. We estimate these values using a sample. The estimated intercept is designated as a and the estimated slope, usually called the **regression coefficient**, is designated as b. The line described by this equation is the line that best "fits" the regression function, and it is called the estimated regression line. Since the line and the equation are one and the same (the equation defines the line), we will be considering them together in the following few sections. There is one point through which the regression line always passes, and that is the point defined by the mean of x and the mean of y $(\bar{x}, \bar{y})$. In figure 14.3 a horizontal line has been placed through this point. A vertical line has been constructed from each value of y to this horizontal line. Each of these vertical lines represents the amount by which the observed value of y deviates from the mean value of y, or

$$y - \bar{y}$$

The sum of these $y - \bar{y}$ values is approximately 0 (table 14.4). However, if we square each of these $y - \bar{y}$ and then sum them, or

$$\Sigma(y - \bar{y})^2$$

we would have a sum of squares for y (recall section 4.3.3.). Note that this value is quite large (table 14.4).

Before terminal confusion sets in, we will review what we have just done here. In effect, we have calculated a sum of squares for y without tak-

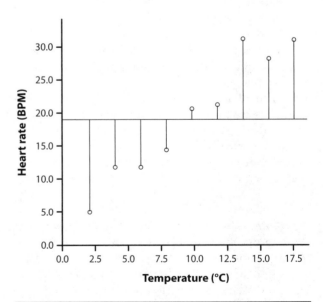

Figure 14.3 Variation in dependent variable when independent variable is not considered.

Table 14.4 Computation of Deviations and Sum of Squared Deviations. Data from table 14.3.

x	y	$y - \bar{y}$	$(y - \bar{y})^2$	$\hat{y}$	$y - \hat{y}$
2	5	−14.89	221.71	5.69	−0.69
4	11	−8.89	79.03	9.24	1.76
6	11	−8.89	79.03	12.79	−1.79
8	14	−5.89	34.69	16.33	−2.34
10	22	2.11	4.45	19.89	2.11
12	23	3.11	9.67	23.44	−0.44
14	32	12.11	146.65	26.99	5.01
16	29	9.11	82.99	30.54	−1.54
18	32	12.11	146.65	34.09	−2.09
Sums: 90	179	−0.01	804.87	$\sum(y - \hat{y})^2 = 48.74$	
Means: 10	19.89				

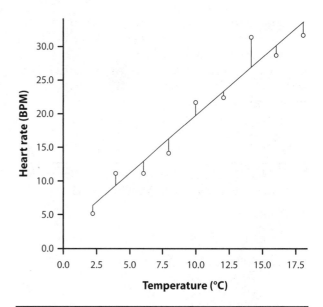

Figure 14.4 Variation in dependent variable when independent variable is considered.

ing x (temperature) into account. In other words, we can see that there is a great deal of variance in heart rate among our nine subjects when we do not consider their temperature.

Suppose now that we could rotate the line in figure 14.3, using $(\bar{x}, \bar{y})$ as a pivot, until it is in a position such that the deviations of the y values from this line were minimized, or more specifically, in a position such that the sum of the squares of the deviations of the y values from the line were minimized (fig. 14.4). Such a line would best "fit" our data. (By the way, this is where the term "**least squares**" regression comes from!) You will note that the observed values of y do not all fall on this line (sometimes none of them do). This is the regression line:

$$\hat{y} = a + bx \qquad (14.3)$$

That value of y that would fall exactly on this line, as described by the regression equation, is referred to as $\hat{y}$ ("y hat"). As before, we have constructed vertical lines from each value of y to this new line, and these vertical lines denote the **residuals**: $e = y - \hat{y}$. These values, given in table 14.4, represent the amount by which each observed value of y deviates from the regression line. For the moment do not be concerned with how we defined the regression line or how the values of $\hat{y}$ were calculated. We will return to this later. Squaring the values of $y - \hat{y}$ in table 14.4 and then summing them gives us another sum of squares for y. This sum of squares represents the variation in y when temperature (x) is taken into account. Note that it is much smaller than the variation in y when x was not considered. In statistical jargon, we have decreased the uncertainty of y by considering x. What does this mean exactly?

If we were given information regarding the heartbeat rate of the nine juniper pythons (example 14.1), but had no knowledge of their temperatures, and if from these data we were asked to predict something about the heart rate of juniper pythons, we could only conclude that the mean heartbeat is around 19.89 BPM with a great deal of variation from animal to animal (estimated from $\sum(y - \bar{y})^2$, which is large). On the other hand, given a knowledge of x (temperature), we could make a much more accurate prediction of the heart rate of juniper pythons.

Refer once again to table 14.4 and figure 14.4 and notice again that the values of y do not exactly coincide with the values of $\hat{y}$. The residuals represent the amount by which y and $\hat{y}$ differ $(y - \hat{y})$. One of the assumptions of regression is that the residuals are normally distributed with a mean of zero (assumption 2, in part).

14.3 Calculating the Estimated Regression Equation

Now we are ready for some formulas, used to estimate the regression parameters. The estimated **slope** of the regression equation is given by

$$b = \frac{\sum xy - \dfrac{\sum x \sum y}{n}}{\sum x^2 - \dfrac{(\sum x)^2}{n}} \qquad (14.4)$$

To obtain this equation, statisticians must solve some simultaneous equations. We'll leave out those details here, although interested readers may consult Neter et al. (1989). Since the least squares regression line goes through the point $(\bar{x}, \bar{y})$, the estimated **y-intercept** is given by

$$a = \bar{y} - b\bar{x} \qquad (14.5)$$

Table 14.5 Intermediate Calculations for Regression Statistics. Data from table 14.3.

	x	y	xy	x^2	y^2
	2	5	10	4	25
	4	11	44	16	121
	6	11	66	36	121
	8	14	112	64	196
	10	22	220	100	484
	12	23	276	144	529
	14	32	448	196	1024
	16	29	464	256	841
	18	32	576	324	1024
sums	90	179	2216	1140	4365

$\Sigma x = 90$ $\Sigma y = 179$ $\Sigma(y-\bar{y})^2 = 804.89$

$\bar{x} = 10$ $\bar{y} = 19.89$ $\Sigma(y-\bar{y})^2 = 48.74$

$\Sigma x^2 = 1140$ $\Sigma y^2 = 4365$

$(\Sigma x)^2 = 8100$ $(\Sigma y)^2 = 32041$

$\Sigma xy = 2216$ $\Sigma x\Sigma y = 16110$

For the juniper python example (table 14.5),

$$b = \frac{2216 - \dfrac{90 * 179}{9}}{1140 - \dfrac{(90)^2}{9}} = 1.775$$

and

$$a = 19.89 - (1.775 * 10) = 2.14$$

The estimated regression function is therefore

$$\hat{y} = 2.14 + 1.775x$$

The regression line is defined by the values of $\hat{y}$ corresponding to values of x. These values are shown in table 14.6. The regression line may be constructed graphically by plotting the values of $\hat{y}$ versus the corresponding values of x. We have now obtained estimates of the slope of the regression equation (b), the y-intercept (a), and the regression line. We now need to attach some statistical significance to these estimates. Table 14.5 above provides the intermediate statistics needed for these statistical tests. Work through the math with your calculator and/or spreadsheet to verify these numbers.

Table 14.6 The Effect of Temperature on the Heart Rate of Juniper Pythons

Temperature (°C) (x)	Heart Rate (y)	$\hat{y}$	e (Residual)
2	5	5.69	–0.69
4	11	9.24	1.76
6	11	12.79	–1.79
8	14	16.33	–2.34
10	22	19.89	2.11
12	23	23.44	–0.44
14	32	26.99	5.01
16	29	30.54	–1.54
18	32	34.09	–2.09

14.4 Testing the Significance of the Regression Equation

Recall that the sample slope, b, is an estimate of the parametric slope, β. Even if $\beta = 0$, which would indicate no dependence of y on x, we might expect b to occasionally have a nonzero value by chance alone. We therefore test the null hypothesis H_0: $\beta = 0$. We could evaluate this null hypothesis either with ANOVA or a t test. Here we will use ANOVA for this hypothesis test. To conduct the ANOVA, we need three **sums of squares** and their associated **degrees of freedom**:

1. the total sum of squares with $n - 1$ degrees of freedom,

2. the regression sum of squares with 1 degree of freedom, and

3. the error sum of squares with $n - 2$ degrees of freedom.

Table 14.5 gives the values needed to calculate these sums of squares by the most-direct methods, shown below. Alternative approaches, which give the same answers for SS_t and SS_e are shown in table 14.4. Check that table as you work through the example below to see that these answers are equivalent.

The total sum of squares, $\Sigma(y-\bar{y})^2$, is given by

$$SS_t = \Sigma y^2 - \frac{(\Sigma y)^2}{n} \qquad (14.6)$$

This is sum of squares for y when x is not considered. (Notice the parallel with equation 4.4 used earlier for descriptive statistics.) For the current example,

$$SS_t = 4365 - \frac{(179)^2}{9} = 804.89$$

The regression sum of squares is given by

$$SS_r = b * \left(\sum xy - \frac{\sum x \sum y}{n} \right) \quad (14.7)$$

This is the sum of squares for y when x is considered.
    For the example,

$$SS_r = 1.775 * \left(2216 - \frac{90 * 179}{9} \right) = 756.15$$

The error sum of squares is given by

$$SS_e = SS_t - SS_r \quad (14.8)$$

This is the sum of squares of the residuals, and it contributes to the variance in y that is still present when x is considered. In other words, it is the error sum of squares. Refer back to figure 14.4, which illustrates these residuals as deviations of points from the regression line.
    For the example,

$$SS_e = 804.89 - 756.15 = 48.74$$

The ANOVA table is constructed as before (section 11.3).

Table 14.7   ANOVA Table for the Data in table 14.3

Source	SS	df	MS	F
Regression	756.15	1	756.15	108.60
Error	48.74	7	6.96	
Total	804.89	8		

From table A.6, the critical value of F for 1 and 7 df at $\alpha = 0.05$ ($F_{1, 7, (0.05)}$) is 5.59. Since $108.60 > 5.59$, we reject H_0: $\beta = 0$ and conclude that the value of y is dependent on the value of x (i.e., that the regression slope is not zero). Although our table of critical F values does not show values associated with lower probability values, the very large F statistic indicates that the p-value is quite small.

Next let's see how *Minitab* displays these same statistics (table 14.8). Notice that the regression equation is the same as we calculated earlier (section 14.4). Below the equation are statistics relating to the y intercept and the slope. *Minitab* displays a t statistic which evaluates H_0: $\beta = 0$ (no dependence of y on x). The very small p-value indicates this hypothesis is rejected (i.e., the dependence of y on x is very strong). We already saw that dependence by eye in our inspection of the scatterplot (fig. 14.1). The strength of the dependence is reported as "R-Sq" (more about this below—section 14.6). The ANOVA table matches what we calculated above in table 14.7. *Minitab* also reports a very small p-value (< 0.001). "Unusual observations" lists the seventh observation, which has a large residual (also see this in fig. 14.4).

14.5 The Confidence Interval for β

Recall that b is an estimate of β, the regression coefficient (the true population slope). As we have seen for estimating other population parameters, we cannot say with certainty what the exact value

Table 14.8   *Minitab* Output for Simple Linear Regression of Heart Rate by Temperature. Data in table 14.3.

Regression Analysis: HeartRate versus Temperature

```
The regression equation is
HeartRate = 2.14 + 1.77 Temperature

Predictor        Coef     SE Coef         T         P
Constant        2.139       1.917      1.12     0.301
Temperature    1.7750      0.1703     10.42     0.000

S = 2.63869      R-Sq = 93.9%      R-Sq(adj) = 93.1%

Analysis of Variance
Source            DF          SS        MS         F         P
Regression         1      756.15    756.15    108.60     0.000
Residual Error     7       48.74      6.96
Total              8      804.89

Unusual Observations

  Obs    Temperature    HeartRate       Fit    SE Fit    Residual    St Resid
    7           14.0        2.000    26.989     1.113       5.011       2.09R

R denotes an observation with a large standardized residual.
```

of β is; however we can compute a 95% confidence interval for β. This requires first determining the amount of uncertainty in our estimate of the slope. The standard error for b, s_b, is given by

$$s_b = \sqrt{\frac{MS_e}{\sum x^2 - \frac{[\sum x]^2}{n}}} \quad \textbf{(14.9)}$$

where MS_e is the mean square for error (from the ANOVA table) and the denominator is the sum of squares for x (equation 4.4). The 95% confidence interval for β is then

$$b \pm s_b(t_{0.05, n-2}) \quad \textbf{(14.10)}$$

where t is the two-tailed probability of t at alpha = 0.05 with $n - 2$ degrees of freedom. For our example, $t_{7, 0.05} = 2.365$. Thus,

$$s_b = \sqrt{\frac{6.96}{1140 - \frac{8100}{9}}} = 0.1703$$

Notice that this statistic is also reported by *Minitab* above (table 14.8). Then, the 95% CI for β is:

$$1.775 \pm (0.1703*2.365)$$
$$= 1.775 \pm 0.403$$

The lower and upper confidence interval for β is thus 1.372 to 2.178.

14.6 The Coefficient of Determination (r^2)

We know that the variance in y is greatly reduced by a knowledge of x, but there is usually still some variance remaining in y when x has been considered (the error variance). If the value of y were completely dependent on x, there would be no error variance, which is to say that all of our observations would fall on the regression line. A general question is, "What proportion of the variance in y is explained by its dependence on x?" To determine this, we compute the **coefficient of determination**, r^2. This value is computed by

$$r^2 = \frac{SS_r}{SS_t} \quad \textbf{(14.11)}$$

For the example,

$$r^2 = \frac{756.15}{804.89} = 0.939$$

Thus, we may conclude that 0.939 or 93.9% of the variance in y is dependent on x, which is to say that

when we know the value of x, we reduce the uncertainty about y by 93.9%. There is still a residual or "unexplained" variance of $100 - 93.9 = 6.1\%$ of the variance in y that is still not explained. This is the variance among individuals that is not related to x. If all the points fell exactly on a straight line, $r^2 = 100\%$ and the unexplained variance would be zero.

The coefficient of determination may also be computed for a correlation analysis (chapter 13), in which case it is the square of the correlation coefficient, r. However, in correlation treatments, r^2 should not be considered as a measure of the variation in y that is explained or dependent upon x, but rather as the variation in y that is *associated* with the variance of x, and vice versa.

14.7 Predicting y from x

One important use of regression is to enable us to predict a value of y for a given value of x. Simply plug a number of x into the equation or find the value of y off the graph. Such predictions must be done with some restraint, however. In the juniper python example, we measured heart rate (y) for temperatures (x) between 2°C and 18°C. Predictions about heart rate much beyond these measured values should be avoided, since we run the risk of predicting nonsense. For example, we would predict a very high heart rate at 100°C by using the regres-

Caution

Although in mathematics, a line extends forever in both directions, in statistics, the line (segment) should only extend over the range of the data (minimum x to maximum x) and no further. Extending the line further in either direction would not be correct, because we do not know if the functional relationship continues to be linear beyond the points we observed. Extending the regression line beyond the range of the points used to generate it is called **extrapolation**, which should usually be avoided. There are times when extrapolation is used, such as in toxicology research. However, interpretation can be very tricky and rests on making some assumptions. For example, see figure 14.5. This figure illustrates a hypothetical toxicology experiment on mice, interpreted to doses to which humans are exposed in the environment. In order to see an effect with a reasonable number of mice (we have room for only so many cages!), the experiment must use high doses. Now, assuming that humans have the same sensitivity as mice and that the function is similar at low doses (continues to be linear), we estimate that a dose of 0.0002 ppm would lead to a cancer risk of 0.000022 (about 22 cases per million). Toxicologists argue a lot about these kinds of experiments.

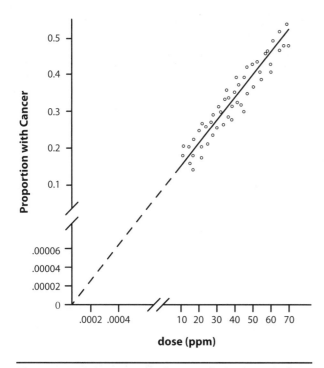

Figure 14.5 An example of extrapolation in toxicology.

sion function (over 179 BPM), but biologically we would probably predict a heart rate of zero (the snakes are dead!) In a similar manner we would probably predict a negative heart rate at temperatures much below 0°C, which is clearly impossible.

Even when predictions about y are kept within reasonable limits of x, it is important to remember that y is usually not an exact function of x, since y is a normally distributed random variable. Thus, when we predict a value of y from x, what we are in fact doing is estimating the population mean value of y for any particular value of x. This estimated value is designated as $\hat{y}$. As usual, we should affix confidence limits to this estimate.

To continue with the juniper python example, we wish to know the mean heart rate of all juniper pythons at 15°C. First, we calculate the value of $\hat{y}$ at $x = 15$ degrees from the regression equation, which is

$$\hat{y} = 2.14 + (1.775 \times 15) = 28.77$$

We now compute the standard error of y, which is given by

$$s_{\hat{y}} = \sqrt{MS_e \left[\frac{1}{n} + \frac{(x - \bar{x})^2}{\sum x^2 - \frac{(\sum x)^2}{n}} \right]} \quad \textbf{(14.12)}$$

where x is that value of x for which we wish to have a confidence interval for y, and $\bar{x}$ is the mean value of x. For the example,

$$s_{\hat{y}} = \sqrt{6.96 \left[\frac{1}{9} + \frac{(15 - 10)^2}{(1140 - \frac{90^2}{9})} \right]} = 1.224$$

The 95% confidence interval for the predicted mean value of y ($\mu_{\hat{y}}$) is:

$$\hat{y} \pm (s_{\hat{y}}) * t_{n-2,(0.05)} \quad \textbf{(14.13)}$$

where t is the critical value of t at alpha 0.05 and $n - 2$ degrees of freedom (2.365). For the example, $t_{7,(0.05)}$ = 2.365 and the 95% CI for $\mu_{\hat{y}}$ is:

$$28.77 \pm (1.224 * 2.365) = 28.77 \pm 2.895$$

In other words, the 95% CI for $\mu_{\hat{y}}$ is 25.87 to 31.67.

We conclude that there is a probability of 0.95 that these limits include the population value of the true heart rate at 15°C. A graphic representation of the confidence interval (or prediction interval) of μ_y at various values of x is shown in figure 14.6. The lines above and below the regression line show the 95% prediction interval for μ_y at any value of x within our measured limits of x. Note that the prediction interval becomes wider at lower and higher values of x. In other words, as we get farther away from the mean of x, our uncertainty in estimating y increases. This trend is because the uncertainty in any estimate of μ_y is due to uncertainty in both the slope and the y-intercept.

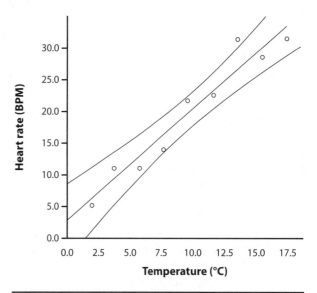

Figure 14.6 The 95% confidence interval (prediction interval) for example 14.1.

14.8 Dealing with Several Values of *y* for Each Value of *x*

Frequently, data that are to be analyzed by regression are conducted as we have outlined previously. Note that in our example, there is only one value of *y* for each value of *x*, which is to say that only one snake was measured at each temperature. For various reasons, it is sometimes desirable or necessary to measure the value of the dependent variable, *y*, for several individuals at each value of the independent variable, *x*. For example, an investigator might wish to test only a few values of *x*. Using only one measurement of *y* at each *x* would result in a sample size that might be too small to reveal a significant association (i.e., a sample size that would result in a high probability of a type II error).

Example 14.3
More Than One Value of *y* for Each Value of *x*

For purposes of illustration, we will redesign the juniper python experiment somewhat, this time using only three temperatures but measuring the heart rate of several snakes at each temperature. The results of the experiment are given in table 14.9.

Table 14.9   The Effect of Temperature on the Heart Rate of Juniper Pythons (Several Values of *y* for Each *x*)

Temperature	Heart Rate
4	9
4	8
4	11
4	8
10	20
10	21
10	19
10	20
10	19
10	20
16	30
16	28
16	31
16	29
16	30

With the data arrayed in this way, we may proceed exactly as before when there was only one value of *y* for each value of *x*. Note carefully, however, that the sum of *x* is not 4 + 10 + 16; rather it is (4 × 4) + (6 × 10) + (5 × 16). In other words, we do not have three values of *x*—we have 15, some of

which are the same! The scatterplot for these data, showing the regression line and the 95% prediction interval, is given in figure 14.7.

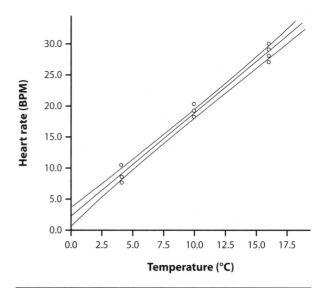

Figure 14.7   The 95% confidence interval for multiple measurements at each value of *x* (example 14.3).

Caution

In the example we have just examined, 15 snakes were used. It is sometimes tempting to measure the same individual repeatedly at each value of *x*, which, in our example, would be only 3 snakes. This is a serious violation of the independence assumption (section 14.1 above). Had we done the experiment in this way, we could use the mean value of *y* at each *x* as a datum (which is sometimes a useful thing to do), but our sample size would be 3, not 15.

14.9 Checking Assumptions and Remedies for Their Failure

As we saw above (section 14.1), for linear regression to work properly several assumptions must be met. The simplest check is to do a scatterplot of the data. This allows us to check for linearity (assumption 5) and also to check for large departures from equal variances (assumption 3). For instance, consider figure 14.8. Panel A illustrates a case where the relationship is strong, but clearly nonlinear. Panel B shows a case where the error variance increases as *x* increases. In both cases, use of linear regression would be inappropriate.

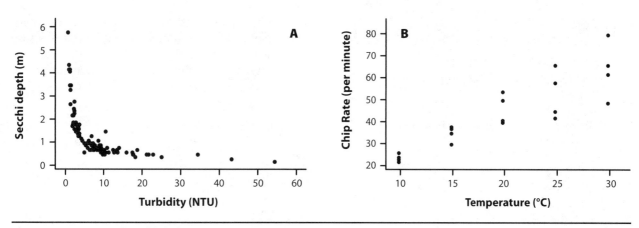

Figure 14.8 Examples of failed regression assumptions.

With statistical packages at our disposal, other plots and more formal tests are available. For instance, an option with the *Minitab* regression analysis routine is to store the residuals. Residuals plots of e against x (or against $\hat{y}$) allow us to more clearly visualize the departure from equal variance. Residuals can be checked for normality (section 11.4). If we know the order of data collection, residuals can also be checked for independence by other procedures.

When the relationship between the variables in a regression analysis does not seem to be linear or the variances are not equal, it is sometimes possible to correct the data by transforming one or both of the variables by methods we saw earlier (see section 11.5).

14.10 Advanced Regression Techniques

In this chapter, you have been introduced to simple linear regression. Regression analysis is a versatile technique and, like ANOVA, has been extended to many different situations. For instance, other mathematical functions may be applied to data that are nonlinear (e.g., exponential, sigmoidal, etc.). Some statistical software can fit functions to these data directly. Alternatively, using the appropriate transformation (e.g. log function on exponential data) allows converting the data to a linear scale, after which linear regression may be performed on the transformed data.

Regression may also be used to model the dependence of y on multiple predictor variables, in a procedure called **Multiple Regression**. For instance, we might model the heart rate of our snakes to simultaneous changes in both temperature and

exposure to some environmental chemical. Modelers sometimes call these "response surfaces," because graphs can be visualized in three dimensions.

Another technique that is widely used is to compare regression lines from two or more groups, in a procedure called **Analysis of Covariance** (ANCOVA). A simple graphic example for two groups is shown in figure 14.9. Basically ANCOVA allows us to ask three questions about the data: (1) Does y depend on x? (2) Is the mean of y for one group different from other groups? And (3) Is the response (slope) of y against x different between groups? Students who expect to use regression analysis in their research are encouraged to explore more-advanced texts, such as Neter et al. (1989), listed in the references.

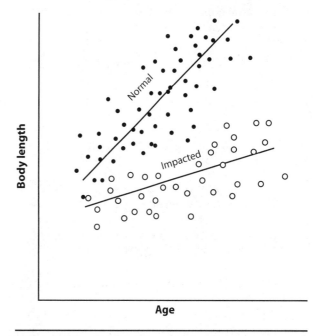

Figure 14.9 Comparing regression lines.

Key Terms

analysis of covariance (ANCOVA)
coefficient of determination (r^2)
dependent variable (= response)
extrapolation
fixed effects (model 1) regression
independent variable (= predictor)
least squares
multiple regression
predictor variable
random effects (model 2) regression
regression analysis
regression coefficient
regression equation
regression line
residual
response variable
simple linear regression
slope
y intercept

Exercises

For each exercise, first graph and inspect a scatterplot of the two variables. Check for any obvious departures from the regression assumptions. Compute the regression equation and coefficient of determination (r^2). Test the null hypothesis that $\beta = 0$. Answer any additional questions specific to each exercise.

14.1 Algae cells were incubated in a culture medium containing different concentrations of dilithium chloride. After a period of incubation, the concentration of dilithium in the algae cells was determined.

Concentration in Medium (μM)	Concentration in Cells (μg/g)
0	0
1	9
2	21
5	47
10	105
20	213

14.2 The cadmium concentrations of grasses at different distances from a major highway were measured, with the following results. (Hint: if the scatterplot does not appear linear, try a transformation to correct the data.)

14.3 Nine male road warblers of different ages were selected and their systolic blood pressure was measured. The results were as follows. After completing the regression analyses, compute the 95% confidence interval for μ_y for each value of x.

Age (years)	Blood Pressure (mm Hg)
1	103
2	115
3	109
4	114
5	120
6	119
7	128
8	132
9	138

14.4 Tomato plants of the same genetic strain and age were subjected to a temperature of 115°F for a period of three hours. Such treatment reduces the plants' ability to photosynthesize. One plant from the group was randomly selected each day for 11 consecutive days, and its rate of photosynthesis was determined. We wish to know if the plants recover from this temperature stress. The results were as follows. After completing the regression analyses, compute the 95% confidence interval for μ_y for each value of x.

Days Posttreatment	Photosynthetic Rate (μM CO_2 g^{-1} s^{-1})
0	15.0
1	17.5
2	16.5
3	19.0
4	22.0
5	24.0
6	22.5
7	26.5
8	25.0
9	30.0
10	29.0

14.5 Bacteria from lake sediments were cultured in the presence of various concentrations of methyl mercury. The number of density of bacteria cells in the various cultures was determined after a period of incubation.

Methyl Mercury (μM)	Cells $\times 10^6$/ml
0	6.6
0	6.9
0	7.2
1	6.8
1	6.0
1	5.6
2	6.4
2	6.0
2	5.4
4	4.8
4	4.4
4	3.9
6	2.6
6	3.1
6	3.4
8	1.0
8	1.3
8	1.7
10	0.2
10	0.3
10	0.5

14.6 Kingfisher nestlings 5–7 days old were randomly assigned, then incubated at different ambient temperatures and their body temperatures measured. At this age, the birds are not capable of maintaining a constant body temperature as are older birds. Does body temperature tend to track environmental temperature? If so, what is the relationship?

Environmental Temperature (°C)	Body Temperature (°C)
10	10
10	12
10	11
10	17
10	15
10	13
20	21
20	28
20	27
20	27
20	24
20	25
20	20
30	29
30	31
30	28
30	35
30	36
30	30
40	40
40	39
40	35
40	37
40	37
40	40
40	38

Data from M. Hamas

14.7 The rate of reaction catalyzed by the enzyme *o*-diphenol oxidase was measured at different substrate concentrations. Rate is the dependent variable. A reciprocal transformation of both variables is required to make the relationship linear. Try plotting both the original data and the transformed data, and note the difference in appearance. Then do regression analysis using the transformed variables.

Substrate Concentration (mM)	Rate (μ1 O_2/min)
0.20	105.3
0.30	142.9
0.40	166.7
0.50	181.8
1.00	256.4

14.8 Ten colonies of juvenile hamsters were established with different densities of animals ranging from 1–5 animals per square meter. After one week, three animals were randomly selected from each colony and their serum corticosterone concentrations (ng/ml) were measured. The relationship does not appear to be linear. Try one or more transformations (see section 11.5) to correct the situation.

Density	Serum Corticosterone
1	3.2
1	2.8
1	3.1
2	8.5
2	10.2
2	9.9
3	27.5
3	34.0
3	29.8
4	97.2
4	120.0
4	105.6
5	330.0
5	285.0
5	315.5

Experimental Design and Selecting Appropriate Procedures

If you are like most students of biostatistics, you probably feel like you have "gone through the wringer." Statistics deals with numerical data and sometimes this seems less like biology and more like math. But we must not lose sight of the primary purpose of this ordeal: to use data to help answer interesting questions. Statistics allows us to make inferences more efficiently and in a way that other scientists generally accept. In this book, you have been introduced to hypothesis testing with a variety of statistical procedures that are appropriate for different sorts of data.

The purpose of this final chapter is to tie these concepts together. We will first review some general principles of experimental design that have been introduced in previous chapters. We will also review the conditions for selecting particular statistical tests we have used so far. Finally, we will briefly explore some newer and specialized statistical procedures.

15.1 Experimental Design

Recall the general purposes to which we have used statistics. Descriptive statistics summarize large masses of data into simpler morsels more easily digested by the researcher or by the reader. Inferential statistics are used for hypothesis testing (the bulk of this book!). Statistics can also be used to guide experimental design, which improves our ability to design meaningful and efficient studies.

Throughout this book, we have assumed that the data collected were sound and could be used to evaluate hypotheses about the statistical populations of interest. Below we briefly consider some of the statistical issues that researchers must consider to obtain sound data. Note that we have deliberately avoided any consideration of the technical details of working in specific fields; we assume that the molecular biologist can correctly determine gene sequences and enzyme activities and that the ecologist can correctly collect, identify, and count species of interest.

As a journey into the unknown, research is a fascinating process. But research can also be frustrating, particularly when methods do not work or when we discover that we spent a lot of effort collecting data that we did not need to collect. In order to avoid some of these pitfalls, it is well worth our time to carefully think about the design for experiments and observational surveys. Let's first consider some general principles for effectively conducting and interpreting experiments (table 15.1 on the following page). These dogmas are distilled from the experience of numerous researchers and statisticians. We discuss several of these principles more fully below.

15.1.1 Observational Surveys Are Useful but Cannot Establish Causation

Survey data are very useful for describing the natural world. How is species diversity related to

Table 15.1   Ten Principles for Sound Experimental Design and Analysis. (Modified after Green 1979)

1. Clearly establish the *hypothesis* being tested and keep the experiment focused on that hypothesis. Be able to explain to someone else the question you are asking.

2. Hypotheses may be tested either with manipulative *experiments* or carefully-focused observational *surveys*. However, determining cause and effect requires controlled experiments.

3. Keep the design as simple as possible. Generally, only a single hypothesis should be addressed with your study. (Exceptions are complex designs, such as two-way ANOVA, that can simultaneously test multiple hypotheses.)

4. Plan to collect *numerical data*. These data may consist of measurements, ranks, or counts of attributes (frequencies).

5. Include a *control*. To test whether a condition has an effect, collect samples both where the condition is present and where the condition is absent. These different *treatment* groups should be alike except for the variable you are trying to test.

6. Beware of *bias*. There is a natural tendency to record what we expect to see, rather than what is actually there. Furthermore, sampling bias occurs when we fail to sample the statistical population we originally had in mind.

7. *Randomize treatments*. Assignment of subjects to different experimental treatments should be done at random. Proper randomization helps to avoid bias and the effects of confounding variables.

8. *Replicate*. Collect multiple measurements from each treatment group. The required sample size depends on both underlying variation and on how small an effect you are trying to detect.

9. Illustrate your data with simple tables and graphs, which show both averages and variation. Differences between groups can be interpreted only in light of variation within groups.

10. Statistical tests can be used to determine whether or not the differences between groups are large enough to consider them truly different.

the area of available habitat? How is heart attack risk related to age, body mass index, and blood pressure? These sorts of studies are important for describing patterns that occur in nature, but they cannot establish causation. This limitation occurs because other hidden variables may be the true causes of the patterns we observe. We could easily go to a number of European cities and count the number of storks and the number of babies born each year and notice a close correlation. Do we then conclude that the storks bring the babies? Or is there another reason for the association we observe? That's why we run experiments.

15.1.2 Avoiding Bias While Testing for Treatment Effects

When our samples are not representative of the population of interest, then our samples are **biased**. We previously described random and biased sampling in section 1.2, as well as methods for collecting random samples (called **randomization**). While important, random sampling is only part of the story. Determining treatment effects in any experiment requires that we reduce other possible sources of bias as well.

To sort out causation we must run a manipulative experiment. But even then, there may be other unmeasured variables that distort the relationship between our variables of interest (**confounding variables**). We could introduce a new vaccine into a population and notice that the incidence of influenza goes down. But the decrease in disease cases

could be due to something completely unrelated to the vaccine. The weather may improve or susceptible individuals may be fewer in number. To sort out confounding variables, every experiment should include a suitable **control**, which is as similar as possible to the treatment of interest. For instance, if human subjects are part of a clinical trial to test a new blood pressure medication, the control might be the most common blood pressure medication on the market. Subjects in each of these two treatments should be matched as closely as possible to every other factor that is known to influence blood pressure (genetics, weight, diet, etc.). To avoid any possible bias in selection of subjects, the particular treatment that a particular subject receives should be assigned at random.

Another source of bias in an experiment is the expectation by people involved in the experiment. We see what we wish to believe. If we test something that we expect to be effective, the researcher, physician, and patient may be influenced in ways that distort the true outcome. For example, if we test a medication for asthma that we hope will improve breathing, then both the researcher and physician may be influenced to see improvement where there is none, and even the patient may breathe easier. For this reason, clinical trials (and other sorts of experiments) should include some form of **blinding**, where the particular treatment that a patient receives is hidden from the doctor diagnosing the ailment. To do this most effectively, a **placebo control** is usually used. A placebo appears in every way to be the same as the treatment of interest, except that it does not include the active ingredient of interest. Many clinical trials now include a **double blind** feature, where neither the physician nor patient knows which treatment is which. Using numbered codes on the bottles, researchers can later determine which patient received which treatment.

The 1954 clinical trial of the Salk polio vaccine provides an excellent example of a good study design. The study involved over 1.8 million children, randomly assigned to vaccine and placebo groups, and used the double blind evaluation of polio symptoms. The study successfully established the effectiveness of Salk's vaccine and the results were widely accepted by the public health community because the study design was sound. For a fascinating description of this study, see the article by Meier (1989) in appendix D.

15.1.3 Reducing Sampling Error

Sampling error represents random "noise" that gets in the way of our seeing the treatment effect, if it indeed exists. **Replication** is one of the most important ways to reduce sampling error. In earlier chapters we emphasized the desirable effects from increasing sample size. Larger sample size (more **replicates**) reduces the sampling error for estimating parameters such as the mean (section 9.2), leads to the distribution of means approaching normality (section 9.3), and increases the power of the test (section 10.6). Recall that the power of the test is the probability of rejecting the null hypothesis when it is indeed false (i.e., the chance of detecting a treatment effect). Following preliminary study, if we know the size of the variance, then we can determine the sample size needed to detect a treatment effect of whatever size we wish (e.g., "Does the new drug cause a 5% reduction in blood pressure?").

For multiple treatments, sampling error is minimized when the replicates are **balanced** among treatments. Thus it should be no surprise that 40 subjects divided into 4 treatments are most useful if we place 10 replicates into each treatment instead of 34 replicates in 1 treatment and 2 replicates each in the other 3 treatments.

Clearly replication is important. But what is a "replicate"? With clinical trials, a single measurement from each human subject would clearly be a replicate. Other times, the replicate is less clear. Consider an example from recent studies of predator-prey ecology. If a researcher investigating the effect of predatory fish chemicals on salamander growth has one aquarium with 20 salamanders receiving the fish cue and another aquarium with 20

salamanders as a control, how many replicates are there per treatment? 20? Surprisingly, the answer is one! The problem is that *the 20 salamanders within a single aquarium are not independent*. There could be a confounding variable related to the aquarium or its pump or the location of the aquarium in the room that influence salamander growth. Furthermore, salamanders in the single aquarium above could very well be interacting through feeding or aggression. Clearly, the measurements of individual salamanders in each tank are not independent. This type of design illustrates a principle called **pseudo-replication**.[1] The superior design would be to use multiple smaller chambers (true replicates), randomly assign treatments to these chambers, and compare the mean growth responses of the salamanders growing in these chambers.

Beginning researchers are well advised to think carefully about their design, considering what truly constitutes a replicate and making this statement explicit in the analysis and report. Since all statistical tests assume that the individual measurements are **independent** of one another, think about your design to be sure that this assumption is met. This is another area where a statistician can provide helpful advice.

Sampling error can also be reduced by recognizing other environmental properties that can be sorted out in sampling or in an experiment. Suppose we are collecting random samples to estimate abundance of aquatic insects that may show preference to different types of aquatic plants. In sampling, we could divide the environment according to plants (say shallow emergent plants vs. submerged plants) and randomly sample within each habitat type. This sort of sampling is called **stratified random sampling**. Similar sampling occurs when an epidemiologist compares the incidence of some disease, say HIV infection, within different groups of people.

This stratified approach can then be employed to sort out environmental variation in an experiment. Consider an experiment with plants in a greenhouse. Even in this carefully-controlled environment the light environment and temperature varies, and thus separate greenhouse benches may be better places to grow than others. If you were conducting a test of four different fertilizers on growth of chrysanthemums, you could put four pots on each of the greenhouse benches. Any variation due to differences in bench location on growth can be later teased out of the analysis. This grouping technique is called **blocking**. In previous chapters we saw experimental designs that employed blocking: paired-comparison tests (section 10.1) and randomized-block ANOVA (section 12.3).

Experimental design is presented more thoroughly in several outstanding references (Box et al. 1978, Mead 1992, Quinn and Keough 2002, Whitlock and Schluter 2008). Students who plan to conduct research in the future are advised to read more thoroughly in this subject area.

15.2 Choosing the Appropriate Statistical Test

Recall that the selection of a statistical test depends on a number of things. What is the hypothesis of interest? What type of variable(s) were measured or counted (section 2.1)? Be able to list and describe the variables and, if categorical, how many different levels of each were selected. What is the structure of the data? Specifically, how many populations were sampled? Are samples related (sections 10.2 and 12.3)? Are the assumptions of parametric statistics met (section 9.8)? What are your sample sizes (numbers of replicates)? Even if you were not doing the analysis yourself, but were instead going to a statistical consultant, (s)he would ask for these kinds of information.

We can use a simple schematic device to aid our selection of statistical tests covered in this book. Recall that we used a shorter version of this key after we covered *t* tests (section 10.8). Figure 15.1 illustrates a key for selecting tests with which you are now familiar. This schematic asks many of the above questions in order to choose the right test. Most often, you will need to read further in a statistics book about the particular test and check examples, to be sure that test applies to your dataset. Please keep in mind that this listing is not exhaustive, as there are other types of statistics which we have not covered in this book. Brief descriptions of some of these procedures are given below in section 15.3 and further information appears in the references.

15.3 A View of Some Other Statistical Procedures

With a limitation of space (and your patience!), we have had to focus our attention in this text on only the most basic and widely-used statistical procedures. A brief description of several other procedures should provide a glimpse of some new possibilities.

Biostatistics is widely used in public health. Besides the traditional statistics covered in this book, public health workers explore **vital statistics** (births, deaths, etc.), which are kept as government records, and can be used to determine such things as age-specific mortality rates and infant mortality rates.[2] Biostatisticians, together with epidemiologists ("disease detectives"), explore the factors related to human disease and death. Understanding the patterns of disease in populations through survey techniques provides the clues to understanding risk factors in circulatory diseases and cancer, as well as tracking the causes of disease outbreaks and epidemics. The role of statistics in public health is described in detail in Lilienfeld and Lilienfeld (1980) and Forthofer and Lee (1995).

Many biological phenomena show patterning in time. For instance, female hormones change with a monthly cycle and the behaviors of numerous creatures vary over a 24-hour period. Detecting the pattern is sometimes easy (graph the data!). Other times it is not. Certain physiological cycles are not so easy to detect and require use of statistical procedures to pull out the chief signal from the "noise" (random variation). If the period of study is long enough, a procedure called **time series analysis** can be used to determine if the cycling is statistically discernible and, if so, what is the period of the response. This method is also applied to long-term studies of the climate and economic cycles. For further information, statistics texts directed toward economists (e.g., Wonnacott and Wonnacott 1977) have detailed descriptions.

Improved computing power has allowed the evolution of **computer-intensive tests**. Examples include **randomization**, **bootstrapping**, and jackknife tests. Computer-intensive procedures are a useful alternative to parametric statistics. These procedures are also useful where large amounts of effort go into obtaining a single estimate and we wish to determine confidence intervals around that estimate. For example, a life table consists of age-structured survivorship and mortality data, which can be used by ecologists to generate a single estimate of population growth rate, which we'll call *r* ("fitness"). To do a bootstrap, we program a computer to randomly discard a single individual from the dataset, use the remaining information to compute *r*, and then replace the individual and repeat the process many times (typically 1000 or more iterations). All these computer-generated *r* values are then used to compute descriptive statistics for *r* (mean and standard deviation) and, from these, a confidence interval. Resampling techniques are described further in Scheiner and Gurevitch (2001) and Whitlock and Schluter (2009), and statistical packages (e.g., *Minitab*) can be readily programmed to carry them out. Guest Box 15.1 by John Heywood on pp. 158–159 provides an introduction to these procedures.

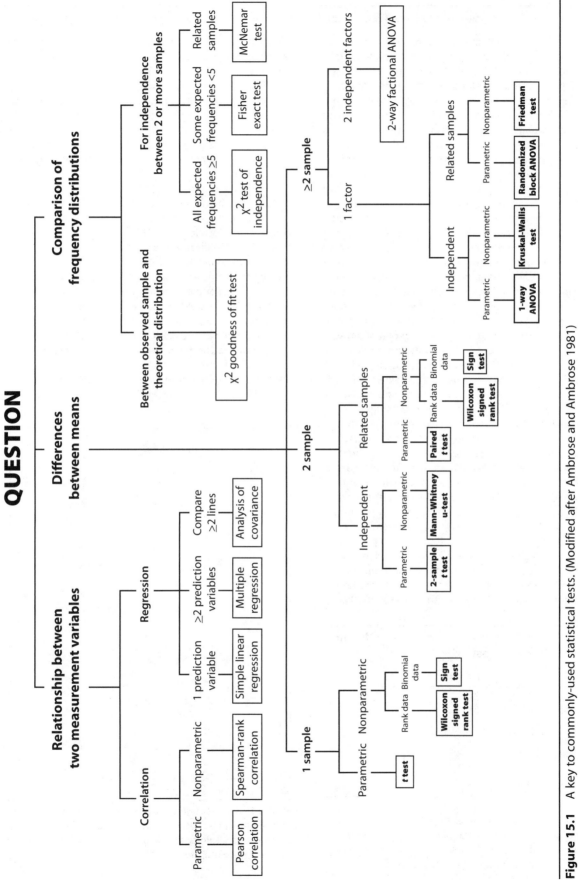

Figure 15.1   A key to commonly-used statistical tests. (Modified after Ambrose and Ambrose 1981)

Guest Box 15.1   Computer-Intensive Techniques *by John Heywood*

With the advent of high speed computers, approaches to statistical testing that require intensive computations have become practical (Manly 2007). A very attractive feature of these **computer-intensive techniques** is that they often require few (or no) assumptions about the distribution from which the sample was obtained. In some cases they allow you to perform statistical tests in situations where it was previously not possible to do so. In other cases they provide assumption-free alternatives to some of the more standard tests. Computer-intensive tests are gaining in popularity and will soon be the standard approach for many types of studies.

Many parameters of interest to biologists are estimated with a statistic for which the sampling distribution is not known. Some examples are the Shannon-Weiner diversity index and the inbreeding coefficient. The distribution of most of these estimators approaches a Normal distribution as sample size becomes large, but statistical inferences are not possible without an estimate of the variance. **Resampling techniques** extract an estimate of the standard error of the estimator by taking a large number of random subsamples from the dataset. In the **bootstrap** method, the original sample of N observations is resampled *with replacement* until N data values are obtained, and this whole process is repeated R times. The sample statistic of interest is calculated for each of these R bootstrap samples, giving a set of R values for the statistic. The mean of these R values is an improved estimate, and their sample standard deviation is the estimated standard deviation (i.e., the standard error) of the bootstrapped estimate. To obtain a precise standard error, a large number of bootstrap samples must be obtained (100 is OK, but 1000 is better). For example, the data in table 1 were obtained from an upland Ozarks forest in southwest Missouri by identifying the emergent tree closest to each of 100 sampling points on a 10×10 square grid (J. S. Heywood, unpublished data). One thousand bootstrap samples from these data generated an estimated Shannon-Weiner diversity index (based on the natural logarithm) of 1.76 with a standard error of 0.05.

Table 1   Tree Species Abundances in an Upland Ozarks Forest

Species	Number
Quercus velutina	25
Quercus falcata	21
Carya tomentosa	18
Quercus alba	14
Quercus stellata	12
Carya texana	8
Quercus marilandica	2

Sometimes a null hypothesis specifies the probability distribution from which a sample was obtained, but the resulting distribution of the test statistic has proved difficult to derive. A computer can generate the unknown distribution by performing a **Monte Carlo simulation**. By taking repeated samples from the known sampling distribution of the data and calculating a value for the test statistic from each sample, the distribution of the statistic can be generated by brute force. For example, we might wish to test the null hypothesis that the 130 trees mapped in figure 1 are located at random within the study plot. The mean nearest-neighbor distance is sensitive to deviations from spatial randomness, so the observed mean nearest neighbor distance in the sample $\left(\overline{x}_O\right)$ is a useful test statistic. A Monte Carlo test would be conducted by programming a computer to generate 130 random locations within the plot and compute the mean nearest-neighbor distance among those locations, $\overline{x}_S$. This simulation is repeated R times ($R \geq 1000$ ideally) and the number of times (k) that $\overline{x}_S > \overline{x}_O$ is recorded. The p-value for a two-tailed test would then be calculated as $p = 2(1 + m)/R$ where m is the smaller of k and $R - k$. For the data in figure 1, only 2 of 10^5 Monte Carlo simulations yielded a value for $\overline{x}_S$ that was less than $\overline{x}_O$, resulting in a p-value of 6×10^{-5} and suggesting that trees are significantly clumped within the study plot.

Randomization tests are probably the most useful of the computer-intensive techniques for statistical inference. A randomization test can be constructed any time the null hypothesis can be rewritten to state that the num-

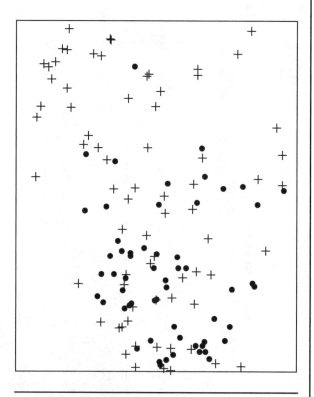

Figure 1   Distribution of 72 post oaks (*Quercus stellata*, +) and 58 black hickories (*Carya texana*, •) in a 60 m × 80 m plot located within an Ozarks savanna (S. H. King and J. S. Heywood, unpublished data).

bers in the dataset are randomly arranged in some fashion. If the null hypothesis is true, then the actual arrangement of the data should be typical of what would be obtained if they were randomly rearranged. If the actual data are arranged in a way that is very different from what is typical for randomly arranged data, then the null hypothesis is rejected. For example, randomization can be used to test the null hypothesis that the oak and hickory trees in figure 1 have the same spatial distribution. Under this null hypothesis, species identities (72 oak and 58 hickory) are randomly assigned to the 130 map locations. For the actual data shown in figure 1, 53 of 89 nearest-neighbor pairs are conspecifics (two oaks or two hickories). When species identities were randomly reassigned to trees 10^6 times, the number of conspecific nearest-neighbor pairs equaled or exceeded 53 only 46,885 times, yielding a one-tailed p-value of 0.0469. This suggests that the oaks and hickories are partially segregated in space within the study plot. Note that a one-tailed test is required since a deficiency of conspecific neighbors would indicate a spatial association between oaks and hickories in excess of what would be generated by their shared spatial distribution.

Computer-intensive procedures can be added to *Minitab* by utilizing its macro programming language. Table 2 shows a *Minitab* macro program that utilizes the data in Table 1 to generate a bootstrapped estimate of the Shannon-Weiner diversity index.

Literature Cited

Manly, Bryan F. J. *Randomization, Bootstrap and Monte Carlo Methods in Biology*, 3rd ed. Boca Raton, FL: Chapman & Hall/CRC, 2007.

Table 2   A *Minitab* Macro that Bootstraps an Estimate of the Shannon-Weiner Diversity Index

```
Gmacro
Shannon
Mtitle "Note:"
Note C2 must contain the number of sampled individuals of each species.
Note Press CONTROL+BREAK to abort.
Mtitle "Shannon-Weiner species diversity index"
Note thinking . . .
Brief 0
Erase c3-c11
Name c4 'Pr'
Name c5 'sample'
Name c11 'S-W Diversity Index'
Name k4 'Bootstrapped estimate'
Name k5 'Bootstrapped standard error'
Let k2=sum(c2)
Let k3=N(c2)
Set c3
    1(1:k3/1)1
End.
Let c4=c2/k2
Do k1=1:1000
    Let c5(1)=k1
    Erase c6-c10
    Random K2 C6;
        Discrete C3 C4.
    Tally c6;
        Counts;
        Store c7 - c8.
    Let c9=c8/k2
    Let c10=-c9*loge(C9)
    Let c11(k1)=sum(c10)
Enddo
Let k4=mean(c11)
Let k5=stdev(c11)
Brief 1
Print k4 k5
Endmacro
```

With the current explosion in computing power, graphic-intensive programs have recently become possible (as any Steven Spielberg movie fan would know). This use of graphics has had important implications for spatial analysis. **Geographic Information Systems** (GIS) are now widely used to organize and display spatially-structured data. Hence GIS is an important tool for data description. GIS also allows calculation of a variety of geographic measures which can be used for many purposes in business, government agencies, and scientists. For instance, stream ecologists and planners both use information on the risk of flooding in river floodplains. Various statistics specific to spatially-structured data (e.g., Mantel tests) are presented in Scheiner and Gurevitch (2001). Other questions dealing with frequency data applied to direction (and time of day) can be analyzed with circular statistics (Zar 1999). For instance, we might want to know from what direction most storms come or which directions birds tend to migrate.

Multivariate statistics are widely used in the fields of ecology and systematics, as well as in the social sciences. These procedures reduce a larger number of response variables into a smaller set of variables which capture the most important features of variation in the dataset and can be shown in plots. Examples of multivariate procedures widely used in biology include Principal Components Analysis (PCA), Canonical Correspondence Analysis (CCA), Multivariate ANOVA, Discriminant Function Analysis, and Cluster Analysis. For descriptions and examples of these tests, see Stevens (1996), McGarigal et al. (2000), or one of the many other books published on multivariate statistics. Most widely-used computer statistics packages (such as *Minitab*, SPSS, SAS) include programs for multivariate analysis, although not all packages contain all the procedures. Several packages are dedicated solely to multivariate analysis (e.g., CANOCO, PC-ORD). Refer to Guest Box 15.2 (on pp. 161–163) for Kim Medley's introduction to multivariate statistics.

Finally, statistics can be used to guide data reviews in areas where extensive research has been previously conducted. For instance, is there a general consensus that compound X is linked to cancer? Occasionally the public is misinformed by the media on conflicts between the results of different scientific studies. (The media, after all, thrive on perceived conflict!) We know from basic design principles (section 15.1 above) that certain key mistakes can invalidate the results of a study (e.g., failure to collect random samples). We also know from the principles of statistics that larger sample size improves estimation of parameters (section 7.3) and the power of statistical tests (section 9.3). Therefore, studies differing in sample sizes can differ in their information content. Also, we may occasionally get a mistaken decision about a hypothesis, even from a well-designed study, because of the nature of samples (recall type 1 and type 2 errors). Formal literature reviews attempt to determine what the general consensus is from a large number of studies. Poorly designed studies can be filtered out and then the remainder subjected to a **meta-analysis**. A meta-analysis is essentially the process of generating statistics about statistics. Overall, is there a statistically discernible effect of compound X? If so, what is the minimum effect size? Meta-analysis allows us to explore subjects which have been extensively studied and argued for many years, and provides a way to contribute toward a general consensus. An entry into literature on meta-analysis can be found in Scheiner and Gurevitch (2001). Refer to Guest Box 15.3 on pp. 164–165 for Jon Shurin's introduction to meta-analysis.

Guest Box 15.2 An Introduction to Multivariate Analysis *by Kim Medley*

Univariate analytical techniques measure associations between one independent variable and a single dependent variable. However, not all research questions lend themselves to a univariate approach. Some sample units have multiple attributes that must be analyzed together. For instance, community ecologists often study associations between biological communities (multiple species co-existing in one area) and characteristics of the abiotic environment. The dependent variables may consist of the number of individuals (i.e., abundance) for each species in a community, and the community may interact with multiple aspects of the abiotic environment (independent variables; e.g., temperature, precipitation, nutrient quantities, soil type, etc.). Because there are as many dependent variables as there are species and there are multiple abiotic variables of interest, a simple linear regression doesn't evaluate the association between the multivariate abiotic environment and the community as a whole. To tackle this problem, community ecologists can use multivariate statistical techniques to: (1) summarize complex relationships by reducing multiple variables to fewer composite variables; (2) evaluate associations between all dependent and independent variables of interest; and (3) define groups (e.g., communities). Indeed these are three major utilities for a multivariate approach. Below is a brief overview of some common techniques within each of these main utilities, with examples for select techniques using a hypothetical mosquito community dataset. These data include mosquito species abundances (table 1) and summaries of important climatic variables (table 2) for 11 US cities.

Simplifying complex datasets. Data reduction is useful to summarize a dataset; averages and frequency distributions (chapters 3 and 4) are forms of data reduction for univariate datasets. Ordination is a data reduction technique used to summarize and visualize complex multivariate relationships. Ordination reduces multiple variables into fewer (usually up to 3) continuous variables along syn-

thetic axes. Ordination plots usually represent sample sites plotted along 2–3 axes based upon axis scores derived from the ordination; axis scores are essentially the x, y, or z axis location for each site. Similar sites (matrix rows)[1] are closer to one another and dissimilar sites are further apart. Alternatively, plot points can represent matrix columns (e.g., species), and proximity represents species that are similar in the sites they occupy (i.e., they tend to co-occur at the same sites). Some commonly used ordination techniques are Principal Components Analysis (PCA), Non-metric Multi-dimensional Scaling (NMS), and Canonical Correspondence Analysis (CCA).

Table 2 Climatic Variables for 11 US Cities Corresponding to Hypothetical Mosquito Communities (table 1). Data were obtained from the WORLDCLIM dataset (www.worldclim.org).

City	Minimum temp. for the coldest month (°C)	Precipitation for the driest month (mm)	Precipitation for the wettest month (mm)
Atlanta, GA	−0.3	80	146
Chicago, IL	−9.3	38	99
Knoxville, TN	−3.1	77	133
Memphis, TN	−1.2	73	136
Miami, FL	15.7	47	197
Nashville, TN	−3.8	70	131
New York, NY	−4.5	78	108
Orlando, FL	8.9	49	192
Portland, ME	−11.4	79	126
Richmond, VA	−3.4	77	109
Springfield, MO	−6.3	46	120

Table 1 Abundance (# of Individuals) for 11 Mosquito Species in Hypothetical Communities at 11 US Cities. Climatic data for these cities are provided in table 2.

City	Aedes aegypt	Aedes albopictus	Aedes triseriatus	Anopheles crucians	Culex pipiens	Culex erraticus	Culex nigripalpus	Psorophora cyanescens	Toxorynchites rutulis	Uranotaenia sapphirina	Wyeomyia smithii
Atlanta, GA	16	50	2	4	0	120	18	1	0	180	6
Chicago, IL	0	0	6	0	18	14	0	0	0	20	12
Knoxville, TN	2	120	10	0	6	82	0	1	0	68	8
Memphis, TN	18	94	5	7	27	68	4	18	0	47	14
Miami, FL	120	6	1	6	0	16	2	2	43	51	0
Nashville, TN	1	68	5	8	34	67	0	12	0	79	18
New York, NY	0	48	87	2	4	1	0	0	0	14	19
Orlando, FL	56	48	0	21	0	109	6	0	5	198	6
Portland, ME	0	0	49	0	12	0	0	0	0	0	51
Richmond, VA	0	49	76	9	29	43	0	14	0	228	98
Springfield, MO	0	12	1	14	7	99	0	8	0	57	1

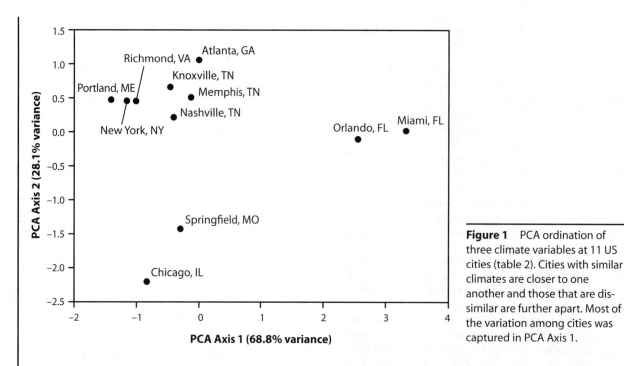

Figure 1 PCA ordination of three climate variables at 11 US cities (table 2). Cities with similar climates are closer to one another and those that are dissimilar are further apart. Most of the variation among cities was captured in PCA Axis 1.

PCA is used when the relationship between variables in the dataset are approximately linear. This is a useful approach when matrices contain non-zero values in most cells, such as the environmental dataset in table 2. The PCA ordination for these data (fig. 1) shows cities as dots positioned along two axes. Cities that have similar climates are closer together and those that are dissimilar are further apart. When a dataset contains a large number of zeros, such as species abundances at multiple sites (e.g., table 1), NMS is a better approach. NMS is also useful to ordinate data that are non-normally distributed. The NMS for mosquito abundances (fig. 2) shows mosquito communities are

quite similar at Memphis, Nashville, and Knoxville, TN, but those communities are quite different from the mosquito community in Portland, ME. PCA and NMS are ordinations using only one matrix, but CCA is a type of ordination that simultaneously considers information in a second matrix—a constrained ordination. CCA attempts to detect structure in the main matrix (e.g., species abundances) by maximizing the relationship with a second matrix (e.g., environmental data). Because of this, CCA plots cannot be used to assess relationships between sites in the same way as PCA and NMS plots. Moreover, CCA is not useful to simply assess structure in the main matrix. However, it is very use-

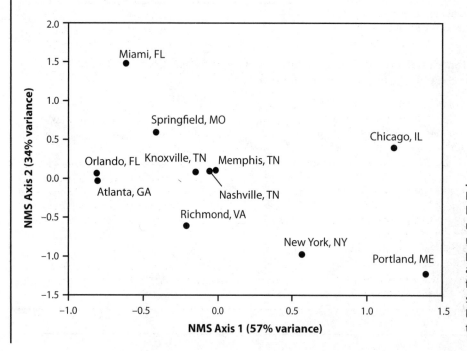

Figure 2 NMS ordination of hypothetical mosquito communities at 11 US cities. Similar mosquito communities are plotted closer to one another and dissimilar communities are further away. Two axes were significant in the NMS, and NMS axis 1 explained more than half of the variance (57%).

ful when the goal is to assess community relationships with environmental variables when the variables included in the second matrix are meaningful (i.e., they must be important to the information in the main matrix).

Measuring associations between variables. Correlation and regression are common univariate approaches to evaluate associations between variables (chapter 13). Multivariate techniques can also test relationships between independent and dependent variables. One widely used approach, the Mantel test, correlates two entire matrices with one another. This correlation technique differs from others presented in this text (chapter 13) because data points in a matrix are inherently non-independent. The Mantel test accounts for this by calculating p-values while accounting for non-independence. Population geneticists commonly use the Mantel test to evaluate the relationship between a genetic distance matrix and a linear distance matrix (i.e., Isolation by Distance; Wright 1943). A significant correlation indicates that as the distance between sample sites increases, genetic distance increases.

Another approach to comparing dependent and independent variables is to reduce a matrix using ordination and compare those results to one or more additional variables. For example, we can test the hypothesis that minimum temperature for the coldest month of the year has a strong effect on mosquito community composition. The matrix containing dependent variables (e.g., species abundances) can be ordinated. Scores (i.e., placement along the axis) for the ordination axis explaining most of the variation in the ordination can then be regressed against environmental variable(s) of interest using simple or multiple regression (see sections 14.1 and 14.10). In this example, the first axis from the NMS shown in figure 2 explained 57% of the variation in the species by site matrix, so we use the scores from this axis as the dependent variable. Multiple regression between climatic variables (table 2) and NMS Axis 1 scores (from fig. 2) revealed mosquito community structure was significantly related to minimum temperature for the coldest month, but not to other measured aspects of climate (table 3). In this way, community structure among cities was summarized by Axis 1 scores, and relationships between communities and climate were statistically evaluated. Similarly, results from PCA ordination can be used to assess the relationship between geographic region (independent variable) and climate (dependent variable). Because climate is a multivariate entity, PCA allows the researcher to reduce climate to an axis score that can be used as the response variable in a simple linear regression with region as the independent (factor) variable.

Defining groups. Ordination plots may reveal groups of sites (or species), but simply observing the plot doesn't reveal whether the groups are significantly more grouped than random. Fortunately, there are techniques to statistically delineate groups of multivariate data. These methods include Cluster Analysis, Multi-Response Permutation Procedures (MRPP), permutation MANOVA (perMANOVA), and Discriminant Analysis (DA). Cluster analysis defines groups

Table 3 ANOVA Table for Multiple Regression Using Aspects of Climate (table 2) as Independent Variables and NMS Axis 1 (from fig. 2) as the Dependent Variable. Temperature: minimum temperature (°C) for the coldest month; precipitation: total precipitation (mm) for the driest month; temperature* precipitation: interaction between temperature and precipitation. Significant association ($p < 0.05$) shown in bold.

Source	df	SS	MS	F	p
temperature	1	3.85	3.85	10.05	**0.02**
precipitation	1	0.17	0.17	0.44	0.53
temperature* precipitation	1	0.70	0.70	1.84	0.22
error	7	2.68	0.38		
total	10				

based upon their similarities and groups them hierarchically; this approach does not require *a priori* prediction of group membership. The results from cluster analysis are displayed in a dendogram, where more similar groups share "nodes." MRPP and perMANOVA are both non-parametric approaches that test significant differences among groups defined before examining the data (*a priori*). MRPP is used for simple designs, whereas perMANOVA is used when designs are more complex. Finally, discriminant analysis determines the relative contribution of variables in the analysis to *a priori* groups.

In summary, multivariate analyses can be very useful for complex, multivariate datasets. This section provided a very brief summary for select methods in three main areas of multivariate analysis. For a more comprehensive look at multivariate methods, students are encouraged to refer to the following sources: Jongman, Ter Braak, & Van Tongeren, 1995; Legendre & Legendre, 1998; McCune & Grace, 2002. Moreover, additional assumptions for each method should be fully researched prior to conducting any of these analyses.

Literature Cited

Jongman, R. H. G., C. J. F. Ter Braak, and O. F. R. Van Tongeren, eds. *Data Analysis in Community and Landscape Ecology*. Cambridge, UK: University Press, 1995.

Legendre, P., & Legendre, L. *Developments in Environmental Modeling*. Amsterdam, the Netherlands: Elsevier Science, 1998.

McCune, B., & Grace, J. B. *Analysis of Ecological Communities*. Gleneden Beach, OR: MjM Software Design, 2002.

Wright, S. "Isolation by Distance." *Genetics* 28, no. 2 (1943): 114–138.

Note

1 Raw multivariate data are organized into matrices; a matrix is a dataset containing data in a row by column format (e.g., tables 1 and 2).

Guest Box 15.3   Meta-Analysis *by Jonathan Shurin*

Most of the statistical analyses presented in this book allow us to interpret the results of an individual study. But what about comparing results across multiple studies to establish whether a pattern is general or a special case? Any single dataset may not represent a broad phenomenon depending on the details of the system studied or how the data were generated. Some questions also lie beyond the scope of a lone investigator, such as those making comparisons across broad geographic scales, habitats or ecosystems. The approach to this problem is called **Meta-analysis**, and the methods used are particularly important in fields like medicine where many studies often address the same question (e.g., the effectiveness or safety of a drug or procedure). The results of these studies may or may not agree, and the differences among them may hold useful information. While using meta-analysis to combine studies increases statistical power, may uncover hidden patterns, and is essential to establishing generality of a particular result, it is also plagued with a number of unique challenges.

In order to draw comparisons across studies, we must decide on a common metric for the effect size of a treatment or association. For manipulative experiments, the effect size measures the contrast between treatment groups in some common currency. One metric is Hedge's "*d*,"

$$d = \frac{\bar{x}_T = \bar{x}_C}{s}$$

where $\bar{x}_T$ is the treatment mean of the dependent variable, $\bar{x}_C$ is the control mean, and s is the pooled standard deviation of the two experimental groups (Hedges & Olkin 1985; Osenberg et al. 1997). Another common metric is the log ratio, R.

$$R = \ln\left(\frac{\bar{x}_T}{\bar{x}_C}\right)$$

Hedge's d represents the difference between the mean values of the control and the treatment relative to the standard deviation within treatments, while R measures the ratio of the two expressed on a log scale. Whether d or R is most appropriate depends on the nature of the data and the question being addressed.

Observational data can also be meta-analyzed. For instance, we might be interested in the slope of the relationship between growth and temperature, body size and metabolism, or species diversity and latitude. Where multiple studies have measured these relationships, their results can be combined in a meta-analysis of regression slopes or strengths of association.

Several issues arise whenever combining data across studies for formal comparison. The first is to decide on the criteria for including studies and evaluating or weighting them based on their quality. Objective criteria for inclusion may include sample size or temporal duration of a study, but there may be issues of data quality apart from these simple metrics. Often the meta-analyzer is forced to make subjective judgments about which studies are most reliable. One way to weight the reliability of different studies is by the within-study variation in the dependent variable. That is, we might have more confidence in studies where the variation within treatment groups is lower than ones where replicates are highly divergent. Hedge's d achieves this by dividing the difference between the treatment and control by the standard deviation, and R can similarly be scaled to the variation by dividing by s. In this case, experiments with high variation among replicates will have less impact on the effect size estimate than those with consistent measures.

Another issue to consider is how to deal with data from experiments that monitor the dependent variable at multiple points over time (Osenberg et al. 1997). For instance, figure 1 shows a hypothetical experiment testing the change in population size of a species grown at two temperatures. The observer sampled the experiment eight times, which is represented by the points on each of the lines. Which time periods should be included in our effect size (d or R)? We will get different answers if we choose samples at times 1, 2 or 3, or the average of all of them. One method is to calculate the rate of change over time by dividing the change in population size by the time span. This works if we use one of the first samples in our data. That is, the high temperature population has higher growth rate than the low temperature population. However, if some experiments last longer than others (e.g., Samples at times 7 vs. 8), than those that run for longer times will have smaller effect sizes because we divide by a longer time. The lesson is that the meta-analyzer needs to carefully study the data (s)he has extracted from the original studies before deciding on an appropriate measure.

Once the correct metric of "effect size" is chosen, each study becomes either a single observation or else several observations if there are multiple contrasts, treatment levels or experiments per study. One can then analyze differences in R and/or d across studies, or associations between

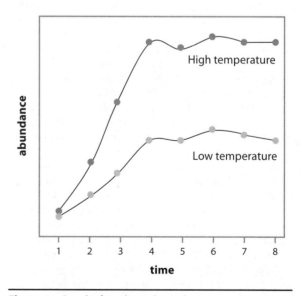

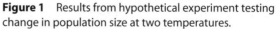

Figure 1   Results from hypothetical experiment testing change in population size at two temperatures.

the effect size metric and other features of the studies. Ecological questions addressed by meta-analysis include:

1. Does species diversity vary in consistent fashion with either the latitude (Hillebrand 2004) or productivity (Mittelbach et al. 2001) of ecosystems?

2. Does plant community productivity respond to either predators (i.e., trophic cascades, Shurin et al. 2002) or nitrogen or phosphorus fertilization (Elser et al. 2007) similarly or differently among ecosystems?

3. How do body size and temperature affect abundance, metabolic rate, species diversity, and rates of evolution of organisms (Damuth 1981; Brown et al. 2004)?

Criticisms of Meta-analysis. Meta-analysis is not accepted by all scientists because the criteria for including studies to compare, and the metrics to use, almost always involve some subjective judgments. For instance, the relationship between diversity and productivity among ecosystems is highly variable, and several meta-analyses have arrived at wildly different conclusions (Whittaker 2010). Indeed, one study using consistent methods in sites across continents found no discernible pattern relating diversity to productivity Adler et al. 2011). These examples show that great caution must be taken in explicitly stating the criteria for including studies, and that detailed analyses must be undertaken to show that the results are not artifacts of the metrics chosen. Nevertheless, meta-analysis fills the vital role of establishing general principles and quantitatively comparing the results of multiple studies to determine the consistency of our findings.

References

Adler, P. B., Seabloom, E. W., Borer, E. T., Hillebrand H., Hautier, Y., Hector, A., Harpole, W. S., O'Halloran, L. R., Grace, J. B., Anderson, T. M., Bakker, J. D., Biederman, L. A., Brown, C. S., Buckley, Y. M., Calabrese, L. B., Chu, C. J., Cleland, E. E., Collins, S. L., Cottingham, K. L., Crawley, M. J., Damschen, E. I., Davies, K. F., DeCrappeo, N. M., Fay, P. A., Firn, J., Frater, P., Gasarch, E. I., Gruner, D. S., Hagenah, N., Lambers, J. H. R., Humphries, H., Jin, V. L., Kay, A. D., Kirkman, K. P., Klein, J. A., Knops, J. M. H., La Pierre, K. J., Lambrinos, J. G., Li, W., MacDougall, A. S., McCulley, R. L., Melbourne, B. A., Mitchell, C. E., Moore, J. L., Morgan, J. W., Mortensen, B., Orrock, J. L., Prober, S. M., Pyke, D. A., Risch, A. C., Schuetz, M., Smith, M. D., Stevens, C. J., Sullivan, L. L., Wang, G., Wragg, P. D., Wright, J. P. & Yang, L. H. "Productivity Is a Poor Predictor of Plant Species Richness." *Science* 333, no. 6050 (2011): 1750–1753.

Brown, J. H., Gillooly, J. F., Allen, A. P., Savage, V. M., & West, G. B. "Toward a Metabolic Theory of Ecology." *Ecology* 85, no. 7 (2004): 1771–1789.

Damuth, J. "Population Density and Body Size in Mammals." *Nature* 290 (April 1981): 699–700.

Elser, J. J., Bracken, M. E. S., Cleland, E. E., Gruner, D. S., Harpole, W. S., Hillebrand, H., Ngai, J. T., Seabloom, E. W., Shurin, J. B., & Smith, J. E. "Global Analysis of Nitrogen and Phosphorus Limitation of Primary Producers in Freshwater, Marine and Terrestrial Ecosystems." *Ecology Letters* 10, no. 12 (2007): 1135–1142.

Hedges, L. V., & Olkin, I. *Statistical Methods for Meta-analysis.* San Diego, CA: Academic Press, 1985.

Hillebrand, H. "On the Generality of the Latitudinal Diversity Gradient." *American Naturalist* 163, no. 2 (2004): 192–211.

Mittelbach, G. G., Steiner, C. F., Scheiner, S. M., Gross, K. L., Reynolds, H. L., Waide, R. B., Willig, M. R., Dodson, S. I., & Gough, L. "What Is the Observed Relationship between Species Richness and Productivity?" *Ecology* 82, no. 9 (2001): 2381–2396.

Osenberg, C. W., Sarnelle, O., & Cooper, S. D. "Effect Size in Ecological Experiments: The Application of Biological Models in Meta-analysis." *American Naturalist* 150, no. 6 (1997): 798–812.

Shurin, J. B., Borer, E. T., Seabloom, E. W., Anderson, K., Blanchette, C. A., Broitman, B., Cooper, S. D., & Halpern, B. S. "A Cross-ecosystem Comparison of the Strength of Trophic Cascades." *Ecology Letters* 5, no. 6 (2002): 785–791.

Whittaker, R. J. "Meta-analyses and Mega-mistakes: Calling Time on Meta-analysis of the Species Richness-productivity Relationship." *Ecology* 91, no. 9 (2010): 2522–2533.

These other statistical tools provide exciting new opportunities for exploring biological data and they are becoming part of our standard vocabulary in statistics. Overall, they have the same usage as the other procedures studied in this book—namely data description, hypothesis testing, and guiding experimental design.

Key Terms

balanced design	meta-analysis
bias	multivariate statistics
blocking	population
bootstrapping	pseudoreplication
computer-intensive tests	random samples
geographic information systems (GIS)	randomization tests
	sample
experimental design	time series analysis
independence	vital statistics

Notes

[1] Pseudoreplication is discussed in detail in: S. H. Hurlbert, "Pseudoreplication and the design of ecological field experiments," *Ecological Monographs* 54 (1984):187–211.

[2] Mortality rate is actually a proportion, commonly expressed as deaths per thousand individuals. Comparison of mortality rates between different populations commonly uses chi-square tests (chapter 7).

Statistical Tables

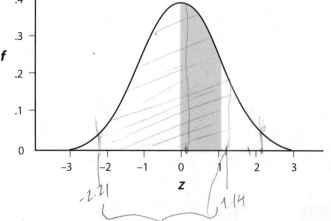

Table A.1 Areas of the Normal Distribution

z	0.00	0.01	0.02	0.03	0.04	0.05	0.06	0.07	0.08	0.09
0.0	0.0000	0.0040	0.0080	0.0120	0.0160	0.0199	0.0239	0.0279	0.0319	0.0359
0.1	0.0398	0.0438	0.0478	0.0517	0.0557	0.0596	0.0636	0.0675	0.0714	0.0754
0.2	0.0793	0.0832	0.0871	0.0910	0.0948	0.0987	0.1026	0.1064	0.1103	0.1141
0.3	0.1179	0.1217	0.1255	0.1293	0.1331	0.1368	0.1406	0.1443	0.1480	0.1517
0.4	0.1554	0.1591	0.1628	0.1664	0.1700	0.1736	0.1772	0.1808	0.1844	0.1879
0.5	0.1915	0.1950	0.1985	0.2019	0.2054	0.2088	0.2123	0.2157	0.2190	0.2224
0.6	0.2258	0.2291	0.2324	0.2357	0.2389	0.2422	0.2454	0.2486	0.2518	0.2549
0.7	0.2580	0.2612	0.2642	0.2673	0.2704	0.2734	0.2764	0.2794	0.2823	0.2852
0.8	0.2881	0.2910	0.2939	0.2967	0.2996	0.3023	0.3051	0.3078	0.3106	0.3133
0.9	0.3159	0.3186	0.3212	0.3238	0.3264	0.3289	0.3315	0.3340	0.3365	0.3389
1.0	0.3413	0.3438	0.3461	0.3485	0.3508	0.3531	0.3554	0.3577	0.3599	0.3621
1.1	0.3643	0.3665	0.3686	0.3708	0.3729	0.3749	0.3770	0.3790	0.3810	0.3830
1.2	0.3849	0.3869	0.3888	0.3907	0.3925	0.3944	0.3962	0.3980	0.3997	0.4015
1.3	0.4032	0.4049	0.4066	0.4082	0.4099	0.4115	0.4131	0.4147	0.4162	0.4177
1.4	0.4192	0.4207	0.4222	0.4236	0.4251	0.4265	0.4279	0.4292	0.4306	0.4319
1.5	0.4332	0.4345	0.4357	0.4370	0.4382	0.4394	0.4406	0.4418	0.4429	0.4441
1.6	0.4452	0.4463	0.4474	0.4484	0.4495	0.4505	0.4515	0.4525	0.4535	0.4545
1.7	0.4554	0.4564	0.4573	0.4582	0.4591	0.4599	0.4608	0.4616	0.4625	0.4633
1.8	0.4641	0.4649	0.4656	0.4664	0.4671	0.4678	0.4686	0.4693	0.4699	0.4706
1.9	0.4713	0.4719	0.4726	0.4732	0.4738	0.4744	0.4750	0.4756	0.4761	0.4767
2.0	0.4772	0.4778	0.4783	0.4788	0.4793	0.4798	0.4803	0.4808	0.4812	0.4817
2.1	0.4821	0.4826	0.4830	0.4834	0.4838	0.4842	0.4846	0.4850	0.4854	0.4857
2.2	0.4861	0.4864	0.4868	0.4871	0.4875	0.4878	0.4881	0.4884	0.4887	0.4890
2.3	0.4893	0.4896	0.4898	0.4901	0.4904	0.4906	0.4909	0.4911	0.4913	0.4916
2.4	0.4918	0.4920	0.4922	0.4925	0.4927	0.4929	0.4931	0.4932	0.4934	0.4936
2.5	0.4938	0.4940	0.4941	0.4943	0.4945	0.4946	0.4948	0.4949	0.4951	0.4952
2.6	0.4953	0.4955	0.4956	0.4957	0.4959	0.4960	0.4961	0.4962	0.4963	0.4964
2.7	0.4965	0.4966	0.4967	0.4968	0.4969	0.4970	0.4971	0.4972	0.4973	0.4974
2.8	0.4974	0.4975	0.4976	0.4977	0.4977	0.4978	0.4979	0.4979	0.4980	0.4981
2.9	0.4981	0.4982	0.4982	0.4983	0.4984	0.4984	0.4985	0.4985	0.4986	0.4986
3.0	0.4987	0.4987	0.4987	0.4988	0.4988	0.4989	0.4989	0.4989	0.4990	0.4990
3.1	0.4990	0.4991	0.4991	0.4991	0.4992	0.4992	0.4992	0.4992	0.4993	0.4993
3.2	0.4993	0.4993	0.4994	0.4994	0.4994	0.4994	0.4994	0.4995	0.4995	0.4995
3.3	0.4995	0.4995	0.4995	0.4996	0.4996	0.4996	0.4996	0.4996	0.4996	0.4997
3.4	0.4997	0.4997	0.4997	0.4997	0.4997	0.4997	0.4997	0.4997	0.4997	0.4998
3.5	0.4998	0.4998	0.4998	0.4998	0.4998	0.4998	0.4998	0.4998	0.4998	0.4998
3.6	0.4998	0.4998	0.4999	0.4999	0.4999	0.4999	0.4999	0.4999	0.4999	0.4999
3.7	0.4999	0.4999	0.4999	0.4999	0.4999	0.4999	0.4999	0.4999	0.4999	0.4999
3.8	0.4999	0.4999	0.4999	0.4999	0.4999	0.4999	0.4999	0.4999	0.4999	0.4999
3.9	0.49995	0.49995	0.49996	0.49996	0.49996	0.49996	0.49996	0.49996	0.49997	0.49997

Table A.2 Critical Values of the *t* Distribution

df	α (Two-Tailed)						
	0.2	**0.1**	**0.05**	**0.02**	**0.01**	**0.001**	**0.0001**
1	3.078	6.314	12.706	31.821	63.657	636.619	6366.198
2	1.886	2.920	4.303	6.695	9.925	31.598	99.992
3	1.638	2.353	3.182	4.541	5.841	12.924	28.000
4	1.533	2.132	2.776	3.747	4.604	8.610	15.544
5	1.476	2.015	2.571	3.365	4.032	6.869	11.178
6	1.44	1.943	2.447	3.143	3.707	5.959	9.082
7	1.415	1.895	2.365	2.998	3.499	5.408	7.885
8	1.397	1.860	2.306	2.896	3.355	5.041	7.120
9	1.383	1.833	2.262	2.821	3.250	4.781	6.594
10	1.372	1.812	2.228	2.764	3.169	4.587	6.211
11	1.363	1.796	2.201	2.718	3.106	4.437	5.921
12	1.356	1.782	2.179	2.681	3.055	4.318	5.694
13	1.35	1.771	2.160	2.650	3.012	4.221	5.513
14	1.345	1.761	2.145	2.624	2.977	4.140	5.363
15	1.341	1.753	2.131	2.602	2.947	4.073	5.239
16	1.337	1.746	2.120	2.583	2.921	4.015	5.134
17	1.333	1.740	2.110	2.567	2.898	3.965	5.044
18	1.33	1.734	2.101	2.552	2.878	3.922	4.966
19	1.328	1.729	2.093	2.539	2.861	3.883	4.897
20	1.325	1.725	2.086	2.528	2.845	3.850	4.837
21	1.323	1.721	2.080	2.518	2.831	3.819	4.784
22	1.321	1.717	2.074	2.508	2.819	3.792	4.736
23	1.319	1.714	2.069	2.500	2.807	3.767	4.693
24	1.318	1.711	2.064	2.492	2.797	3.745	4.654
25	1.316	1.708	2.060	2.485	2.787	3.725	4.619
26	1.315	1.706	2.056	2.479	2.779	3.707	4.587
27	1.314	1.703	2.052	2.473	2.771	3.690	4.558
28	1.313	1.701	2.048	2.467	2.763	3.674	4.530
29	1.311	1.699	2.045	2.462	2.756	3.659	4.506
30	1.31	1.697	2.042	2.457	2.750	3.646	4.482
40	1.303	1.684	2.021	2.423	2.704	3.551	4.321
60	1.296	1.671	2.000	2.390	2.660	3.460	4.169
100	1.292	1.660	1.984	2.364	2.626	3.390	4.053
∞	1.282	1.645	1.960	2.326	2.576	3.291	3.750

interpolate $\dfrac{2.021 + 2.000}{2} = 2.011$

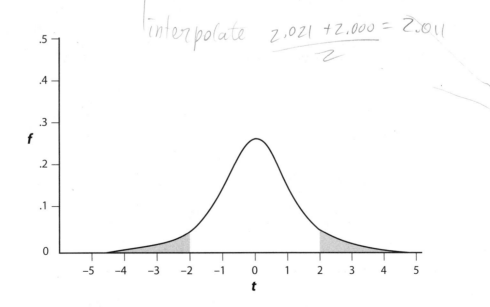

Table A.3   Critical Values of the Chi-Square Distribution

df	α						
	0.50	0.10	0.05	0.02	0.01	0.001	0.0001
1	0.46	2.71	3.84	5.41	6.63	10.83	15.14
2	1.39	4.60	5.99	7.82	9.21	13.82	18.42
3	2.37	6.25	7.81	9.84	11.34	16.27	21.11
4	3.36	7.78	9.49	11.67	13.28	18.47	23.51
5	4.35	9.24	11.07	13.39	15.09	20.51	25.74
6	5.35	10.64	12.59	15.03	16.81	22.46	27.86
7	6.35	12.02	14.07	16.62	18.48	24.32	29.88
8	7.34	13.36	15.51	18.17	20.09	26.12	31.83
9	8.34	14.68	16.92	19.68	21.67	27.88	33.72
10	9.34	15.99	18.31	21.16	23.21	29.59	35.56
11	10.34	17.28	19.68	22.62	24.72	31.26	37.37
12	11.34	18.55	21.03	24.05	26.22	32.91	39.13
13	12.34	19.81	22.36	25.47	27.69	34.53	40.87
14	13.34	21.06	23.68	26.87	29.14	36.12	42.58
15	14.34	22.31	25.00	28.26	30.58	37.70	44.26
16	15.34	23.54	26.30	29.63	32.00	39.25	45.92
17	16.34	24.77	27.59	31.00	33.41	40.79	47.57
18	17.34	25.99	28.87	32.35	34.81	42.31	49.19
19	18.34	27.20	30.14	33.69	36.19	43.82	50.80
20	19.34	28.41	31.41	35.02	37.57	45.31	52.39
21	20.34	29.62	32.67	36.34	38.93	46.80	53.96
22	21.34	30.81	33.92	37.66	40.29	48.27	55.52
23	22.34	32.01	35.17	38.97	41.64	49.73	57.08
24	23.34	33.20	36.42	40.27	42.98	51.18	58.61
25	24.34	34.38	37.65	41.57	44.31	52.62	60.14
26	25.34	35.56	38.89	42.86	45.64	54.05	61.66
27	26.34	36.74	40.11	44.14	46.96	55.48	63.16
28	27.34	37.92	41.34	45.42	48.28	56.89	64.66
29	28.34	39.09	42.56	46.69	49.59	58.30	66.15
30	29.34	40.26	43.77	47.96	50.89	59.70	67.63

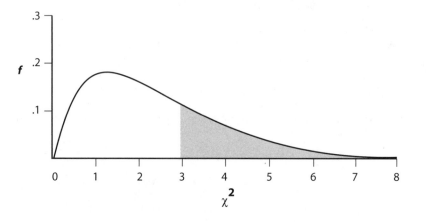

Table A.4 Critical Values of *U* for the Mann-Whitney U Test. Sample sizes for each group are denoted by n_1 and n_2 and which group is which does not matter. At $\alpha = 0.05$, use the top table for a two-tailed test and bottom table for a one-tailed test. (The top table may also be used for a one-tailed test at $\alpha = 0.025$, and the bottom table for a two-tailed test at $\alpha = 0.10$.)

Two-tailed test

5	6	7	8	9	10	11	12	13	14	15	16	17	18	19	20	n_2
23	27	30	34	38	42	46	49	53	57	61	65	68	72	76	80	5
	31	36	40	44	49	53	58	62	67	71	75	80	84	89	93	6
		41	46	51	56	61	66	71	76	81	86	91	96	101	106	7
			51	57	63	69	74	80	86	91	97	102	108	114	119	8
				64	70	76	82	89	95	101	107	114	120	126	132	9
					77	84	91	97	104	111	118	125	132	138	145	10
						91	99	106	114	121	129	136	143	151	158	11
							107	115	123	131	139	147	155	163	171	12
								124	132	141	149	158	167	175	184	13
									141	151	160	169	178	188	197	14
										161	170	180	190	200	210	15
											181	191	202	212	222	16
												202	213	224	235	17
													225	236	248	18
														248	261	19
															273	20

One-tailed test

5	6	7	8	9	10	11	12	13	14	15	16	17	18	19	20	n_2
21																5
25	29															6
29	34	38														7
32	38	43	49													8
36	42	48	54	60												9
39	46	53	60	66	73											10
43	50	58	65	72	79	87										11
47	55	63	70	78	86	94	102									12
50	59	67	76	84	93	101	109	118								13
54	63	72	81	90	99	108	117	126	135							14
57	67	77	87	96	106	115	125	134	144	153						15
61	71	82	92	102	112	122	132	143	153	163	173					16
65	76	86	97	108	119	130	140	151	161	172	183	193				17
68	80	91	103	114	125	137	148	159	170	182	193	204	215			18
72	84	96	108	120	132	144	156	167	179	191	203	214	226	238		19
75	88	101	113	126	138	151	163	176	188	200	213	225	237	250	262	20

Table A.5   **Critical Values of T for the Wilcoxon Signed Rank Test. The row *n* represents the number of matched pairs without ties. Reject H$_0$ if the statistic is *less than or equal to* the critical value. For a 1-tailed test, divide the probabilities at the top of the table by 2.**

	α (2 tailed)		
n	0.10	0.05	0.01
6	2	0	—
7	3	2	—
8	5	3	0
9	8	5	1
10	10	8	3
11	13	10	5
12	17	13	7
13	21	17	9
14	25	21	12
15	30	25	15
16	35	29	19
17	41	34	23
18	47	40	27
19	53	46	32
20	60	52	37
21	67	58	42
22	75	65	48
23	83	73	54
24	91	81	61
25	100	89	68

Table A.6 Critical Values of the F Distribution ($\alpha = 0.05$; df_1 = Treatment Degrees of Freedom, df_2 = Error Degrees of Freedom). Part A: $\alpha = 0.05$

df_2	1	2	3	4	5	6	7	8	9	10	11	12
2	18.5	19.0	19.2	19.3	19.4	19.4	19.4	19.4	19.4	19.4	19.4	19.4
3	10.1	9.55	9.28	9.12	9.01	8.94	8.89	8.85	8.81	8.79	8.76	8.74
4	7.71	6.94	6.59	6.39	6.26	6.16	6.09	6.04	6.00	5.96	5.93	5.91
5	6.61	5.79	5.41	5.19	5.05	4.95	4.88	4.82	4.77	4.74	4.71	4.68
6	5.99	5.14	4.76	4.53	4.39	4.28	4.21	4.15	4.10	4.06	4.03	4.00
7	5.59	4.74	4.35	4.12	3.97	3.87	3.77	3.73	3.68	3.64	3.60	3.57
8	5.32	4.46	4.07	3.84	3.69	3.58	3.50	3.44	3.39	3.35	3.31	3.28
9	5.12	4.26	3.86	3.63	3.48	3.37	3.29	3.23	3.18	3.14	3.10	3.07
10	4.96	4.10	3.71	3.48	3.33	3.22	3.14	3.07	3.02	2.98	2.94	2.91
11	4.84	3.98	3.59	3.36	3.20	3.09	3.01	2.95	2.90	2.85	2.82	2.79
12	4.75	3.89	3.49	3.26	3.11	3.00	2.91	2.85	2.80	2.75	2.72	2.69
15	4.54	3.68	3.29	3.06	2.90	2.79	2.71	2.64	2.59	2.54	2.51	2.48
20	4.35	3.49	3.10	2.87	2.71	2.60	2.51	2.45	2.39	2.35	2.31	2.28
25	4.24	3.39	2.99	2.76	2.60	2.49	2.40	2.34	2.28	2.24	2.21	2.16
30	4.17	3.32	2.92	2.69	2.53	2.42	2.33	2.27	2.21	2.16	2.13	2.09
40	4.08	3.23	2.84	2.61	2.45	2.34	2.25	2.18	2.12	2.08	2.04	2.04
60	4.00	3.15	2.76	2.53	2.37	2.25	2.17	2.10	2.04	1.99	1.95	1.92
120	3.92	3.07	2.68	2.45	2.29	2.17	2.09	2.02	1.96	1.91	1.87	1.83

Table A.6 Part B: $\alpha = 0.01$

df_2	1	2	3	4	5	6	7	8	9	10	11	12
2	98.5	99.0	99.2	99.2	99.3	99.3	99.4	99.4	99.4	99.4	99.4	99.4
3	34.1	30.8	29.5	28.7	28.2	27.9	27.7	27.5	27.3	27.2	27.1	27.1
4	21.2	18.0	16.7	16.0	15.5	15.2	15.0	14.8	14.7	14.5	14.4	14.4
5	16.3	13.3	12.1	11.4	11.0	10.7	10.5	10.3	10.2	10.1	9.99	9.89
6	13.7	10.9	9.78	9.15	8.75	8.47	8.26	8.10	7.98	7.87	7.79	7.72
7	12.2	9.55	8.45	7.85	7.46	7.19	6.99	6.84	6.72	6.62	6.54	6.47
8	11.3	8.65	7.59	7.01	6.63	6.37	6.18	6.03	5.91	5.81	5.73	5.67
9	10.6	8.02	6.99	6.42	6.06	5.80	5.61	5.47	5.35	5.26	5.18	5.11
10	10.0	7.56	6.55	5.99	5.64	5.39	5.20	5.06	4.94	4.85	4.77	4.71
11	9.65	7.21	6.22	5.67	5.32	5.07	4.89	4.74	4.63	4.54	4.46	4.40
12	9.33	6.93	5.95	5.41	5.06	4.82	4.64	4.50	4.39	4.30	4.22	4.16
15	8.68	6.36	5.42	4.89	4.56	4.32	4.14	4.00	3.89	3.80	3.73	3.67
20	8.10	5.85	4.94	4.43	4.10	3.87	3.70	3.56	3.46	3.37	3.29	3.23
25	7.77	5.57	4.68	4.18	3.86	3.63	3.46	3.32	3.22	3.13	3.06	2.99
30	7.56	5.39	4.51	4.02	3.70	3.47	3.30	3.17	3.07	2.98	2.90	2.84
40	7.31	5.18	4.31	3.83	3.51	3.29	3.12	2.99	2.89	2.80	2.73	2.66
60	7.08	4.98	4.13	3.65	3.34	3.12	2.95	2.82	2.72	2.63	2.56	2.50
120	6.85	4.79	3.95	3.48	3.17	2.96	2.79	2.66	2.56	2.47	2.40	2.34

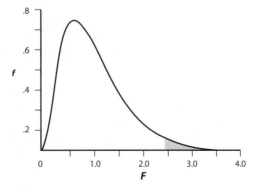

Table A.7   Critical Values of q (Studentized t) for the Tukey Test ($\alpha = 0.05$)

Error df	Number of Groups (Treatments)								
	2	3	4	5	6	7	8	9	10
1	17.97	26.98	32.82	37.08	40.41	43.12	45.40	47.36	49.07
2	6.08	8.33	9.80	10.88	11.74	12.44	13.03	13.54	13.99
3	4.50	5.91	6.82	7.50	8.04	8.48	8.85	9.18	9.46
4	3.93	5.04	5.76	6.29	6.71	7.05	7.35	7.60	7.83
5	3.64	4.60	5.22	5.67	6.03	6.33	6.58	6.80	6.99
6	3.46	4.34	4.90	5.30	5.63	5.90	6.12	6.32	6.49
7	3.34	4.16	4.68	5.06	5.36	5.61	5.82	6.00	6.16
8	3.26	4.04	4.53	4.89	5.17	5.40	5.60	5.77	5.92
9	3.20	3.95	4.41	4.76	5.02	5.24	5.43	5.59	5.74
10	3.15	3.88	4.33	4.65	4.91	5.12	5.30	5.46	5.60
11	3.11	3.82	4.26	4.57	4.82	5.03	5.20	5.35	5.49
12	3.08	3.77	4.20	4.51	4.75	4.95	5.12	5.27	5.39
13	3.06	3.73	4.15	4.45	4.69	4.88	5.05	5.19	5.32
14	3.03	3.70	4.11	4.41	4.64	4.83	4.99	5.13	5.25
15	3.01	3.67	4.08	4.37	4.59	4.78	4.94	5.08	5.20
16	3.00	3.65	4.05	4.33	4.56	4.74	4.90	5.03	5.15
17	2.98	3.63	4.02	4.30	4.52	4.70	4.86	4.99	5.11
18	2.97	3.61	4.00	4.28	4.49	4.67	4.82	4.96	5.07
19	2.96	3.59	3.98	4.25	4.47	4.65	4.79	4.92	5.04
20	2.95	3.58	3.96	4.23	4.45	4.62	4.77	4.90	5.01
24	2.92	3.53	3.90	4.17	4.37	4.54	4.68	4.81	4.92
30	2.89	3.49	3.85	4.10	4.30	4.46	4.60	4.72	4.82
40	2.86	3.44	3.79	4.04	4.23	4.39	4.52	4.63	4.73
60	2.83	3.40	3.74	3.98	4.16	4.31	4.44	4.55	4.65
120	2.80	3.36	3.68	3.92	4.10	4.24	4.36	4.47	4.56
∞	2.77	3.31	3.63	3.86	4.03	4.17	4.29	4.39	4.47

Table A.8   Critical Values of the Pearson Correlation Coefficient (r) (α = 0.05)*

df	r
1	0.997
2	0.950
3	0.878
4	0.811
5	0.754
6	0.707
7	0.666
8	0.632
9	0.602
10	0.576
11	0.553
12	0.532
13	0.514
14	0.497
15	0.482
16	0.468
17	0.456
18	0.444
19	0.433
20	0.423
21	0.413
22	0.404
23	0.396
24	0.388
25	0.381
26	0.374
27	0.367
28	0.361
29	0.355
30	0.349
35	0.325
40	0.304
45	0.288
50	0.273
60	0.250
70	0.232
80	0.217
90	0.205
100	0.195
120	0.174

*For values not in the table, test for significance by the t test (see text, section 13.1).

Table A.9   Critical Values of the Spearman Rank Correlation Coefficient (α = 0.05)*

n	r_s
4	1.000
5	0.900
6	0.829
7	0.714
8	0.643
9	0.600
10	0.564
12	0.506
14	0.456
16	0.425
18	0.399
20	0.377

*For sample sizes not tabulated, test for significance using the t test (see text).

Table A.10   Table of Random Numbers

874335	218040	632420	240295	301131	152740	433058	274170
142131	051859	719342	714391	174251	147150	108520	771712
577728	460401	847722	767239	201744	006565	204589	960553
080052	246887	107893	627841	196599	792021	038162	390011
501153	355165	168311	790826	174928	955178	754258	125025
146207	369709	775557	516449	855970	838321	826020	246163
273515	015616	254341	330587	162088	174360	554720	349616
594504	658609	007492	524747	718771	586831	569750	047201
773722	805035	015969	656055	354632	089893	328631	466358
928848	601866	338853	047266	601409	588331	617007	750155
130680	336701	613351	286758	193966	377556	048648	557283
903145	937763	554796	728537	570290	643603	565449	562057
723294	473898	456644	992231	371495	963132	937428	954420
521302	654580	690478	463092	941820	803428	262731	939938
180471	329905	206005	792002	828627	022402	467626	239803
037226	990598	031055	395463	282404	368588	806509	590830
381118	268005	771588	955604	756766	981147	361899	245461
954822	434100	111684	920179	408451	889864	544440	471762
454139	901479	313550	002567	597321	515148	592903	053426
027996	723365	717520	681773	386364	168036	074181	789768
778443	093607	242049	702424	041696	550187	383294	995730
260656	846676	883719	574775	532552	253887	243386	001878
982935	957671	217239	074705	031298	262045	205728	654403
906706	042314	895439	743718	413420	448197	149714	815122
946521	856953	149277	388942	757533	076503	782862	861477
470054	798560	287835	583131	845375	301748	140819	186534
798107	404733	198320	164665	661808	669342	087352	698984
704605	853694	846064	737547	894822	615321	814358	323143
916600	464292	774523	171407	435529	966344	341855	498953
614267	196000	605281	101497	878168	439697	017987	681981
930906	148913	538043	428698	020102	143290	019025	843417
452944	063756	850643	819512	361819	075658	849363	970079
719931	821876	399037	206069	606933	625961	841521	564408
724544	945246	117307	286123	162181	073984	656142	144469
412582	096463	517660	023052	637428	090138	781997	743955
182972	578750	190428	145861	345662	235457	035980	412182
387765	835955	304068	649179	802995	461602	063111	714091
832135	952549	105163	293258	228666	610859	836534	230248
274385	153632	418418	103979	045038	916136	157518	056846
925940	304925	146667	872845	377600	500970	155459	305700

Table A.11 95% Confidence Intervals for *p* as Estimated by the Sample Proportion $\hat{P}$ for Different Sample Sizes. In each cell, the top number represents the lower confidence limit and the bottom number represents the upper limit. Values from Rohlf and Sokal (2012).

$\hat{P}$	20	50	100	200	500	1000
0.05	.0026	.0119	.0200	.0268	.0338	.0379
	.2442	.1519	.1134	.0893	.0727	.0653
0.10	.0181	.0403	.0534	.0642	.0757	.0824
	.3199	.2176	.1740	.1494	.1297	.1203
0.15	.0422	.0711	.0890	.1060	.1208	.1289
	.3722	.2782	.2340	.2069	.1838	.1734
0.20	.0714	.1066	.1287	.1495	.1668	.1759
	.4235	.3385	.2893	.2620	.2378	.2259
0.25	.1041	.1462	.1741	.1946	.2139	.2240
	.4745	.3891	.3444	.3147	.2898	.2779
0.30	.1396	.1880	.2142	.2396	.2609	.2720
	.5254	.4396	.3996	.3672	.3419	.3294
0.35	.1669	.2281	.2591	.2855	.3089	.3210
	.5764	.4899	.4498	.4198	.3929	.3804
0.40	.2090	.2680	.3056	.3322	.3576	.3700
	.6277	.5402	.5000	.4699	.4439	.4309
0.45	.2443	.3177	.3546	.3824	.4060	.4190
	.6800	.5906	.5501	.5200	.4939	.4809
0.50	.2928	.3589	.3997	.4299	.4560	.4690
	.7072	.6411	.6003	.5071	.5440	.5310

Answers to Odd-Numbered and Selected Exercises

Answers to the exercises starting with chapter 2 are shown below. Rounding errors may cause slight discrepancies between these answers and your answers.

Chapter 2

2.1 a) continuous   b) 177.4 cm

2.3 discrete

2.5

Male ID	1	2	③	4	⑤	6
Height	181	202	190	185	190	200
Rank	1	6	3.5	2	3.5	5

2.7 a) continuous and derived
  b) discrete
  c) attribute (nominal)
  d) continuous, transformed
  e) continuous and derived
  f) rank (ordinal)

$$\frac{3+4}{2} = 3.5$$

Chapter 3

3.1

Height (cm)	f
159.5–161.5	1
161.5–163.5	0
163.5–165.5	0
165.5–167.5	0
167.5–169.5	1
169.5–171.5	4
171.5–173.5	9
173.5–175.5	8
175.5–177.5	13
177.5–179.5	18
179.5–181.5	20
181.5–183.5	18
183.5–185.5	18
185.5–187.5	11
187.5–189.5	7
189.5–191.5	10
191.5–193.5	4
193.5–195.5	2
195.5–197.5	2
197.5–199.5	1
199.5–201.5	1

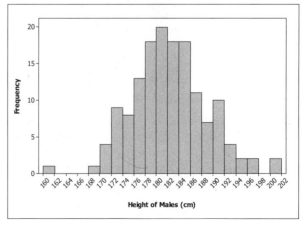

179

3.3

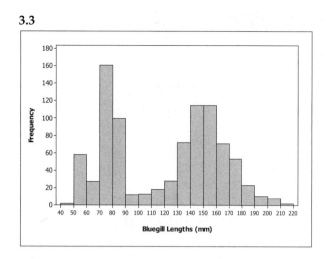

3.5

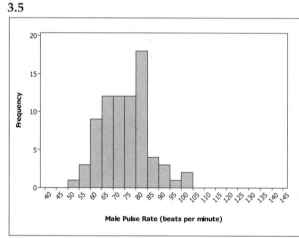

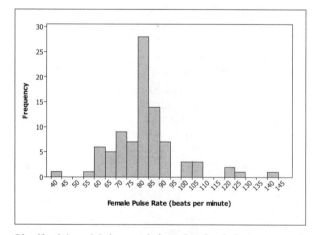

Similarities: Males and females both have a peak between 80–85 beats per minute.
Differences: Females have a larger range of pulse rates (40–145) than males (50–105).

3.7

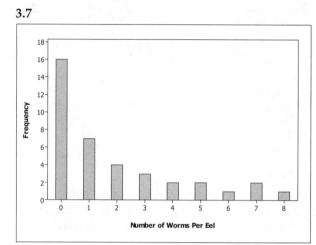

Total number of worms found in 40 eels: 89

3.9

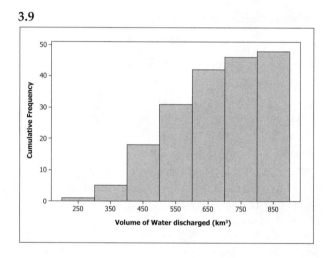

Chapter 4

4.1 Mean 6.0, Median 6.5, Range 1, 11, IQ range 3.0, 9.0, Variance 8.97, Standard Deviation 2.99, CV 49.9%

4.3 Mean 55.70, Median 56.5, Range 47.0, 62.4, IQ range 52.175, 59.225, Variance 18.43, Standard Deviation 4.293, CV 7.708%

4.5 Weighted Mean 7.57

4.7 Median = 2

4.9 Answers will vary.

Chapter 5

5.1 a) P(brown) = 2/5, P(white) = 4/15, P(tan) = 1/3
b) P(brown or tan) = 11/15

5.3 P(2 heads) = 0.25

5.5 P(1 head and 1 tail) = 0.50

5.7 P(2 white) = 28/435

5.9 P(First is A | Second is O) = 3/46

5.11 # of possible combinations = 1330

5.13 $k = 6$ $p = 1/2$ $q = 1/2$
$p(x = 2) = 0.234375$

5.15 $k = 5$ $p = 1/6$ $q = 5/6$
$p(x = 5) = 0.000129$

5.17 $k = 5$ $p = 1/6$ $q = 5/6$
$p(x \leq 2) = 0.964506$

5.19 $k = 6$ $p = 1/2$ $q = 1/2$
$p(x = 3) = 0.3125$

5.21 a) $p(x = 0) = 0.177979$
b) $p(x \geq 1) = 0.822021$ *greater or = 1*

5.23

x	$p(x)$
0	0.177979
1	0.355957
2	0.296631
3	0.131836
4	0.032959
5	0.004395
6	0.000244

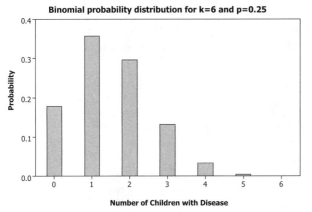

Binomial probability distribution for k=6 and p=0.25

5.25

x	$p(x)$
0	0.972948
1	0.026805
2	0.000246
3	0.000001

$E(x) = p(x)*n = 972.984 \approx 973$

5.27 mean = 2.99 $\Sigma f = 100$

x	$p(x)$
0	0.0503
1	0.1504
2	0.2248
3	0.2240
4	0.1675
5	0.1001
6	0.0499
7	0.0213
8	0.0080

5.29 $\mu = 5/3$ phage/cell
a) $p(x = 0) = 0.188876$
b) $p(x \geq 1) = 0.811124$

Chapter 7

7.1 $x^2 = 25$, $x^2_{1,(.05)} = 3.84$, $p < 0.0001$, reject H_0, yes

7.3 $x^2 = 0.333$, $x^2_{1,(.05)} = 3.84$, $p > 0.5$, fail to reject H_0, does not differ from expected ratio

7.5 $x^2 = 1.712$, $x^2_{4,(.05)} = 9.49$, $p > 0.5$, fail to reject H_0, no significant difference

7.7 (The last categories should be combined, df = $7 - 1 - 1 = 5$.)
$x^2 = 0.2113$, $x^2_{5,(.05)} = 11.07$, $p > 0.50$, fail to reject H_0, distribution is random

7.9 $x^2 = 39.42$, $x^2_{1,(.05)} = 3.84$, $p < 0.0001$, reject H_0, yes there is an association

7.11

	SMK Yes	SMK NO	Total
Male	14	63	77
	11.67	65.33	
	0.467	0.083	
Female	11	77	88
	13.33	74.67	
	0.408	0.073	
Total	25	140	165

Chi-Sq = 1.031, DF = 1, P-Value = 0.310

7.13 Pain Relief: $x^2 = 17.282$, $x^2_{2,(.05)} = 5.99$, $0.0001 < p < 0.001$ reject H_0
Stiffness relief: $x^2 = 11.618$, $x^2_{2,(.05)} = 5.99$, $0.001 < p < 0.01$ reject H_0

7.15 One-tailed $p = p_1 + p_2 = 0.05128$, fail to reject H_0

7.17 Chicken dressing, $x^2 = 11.175$, $x^2_{1,(.05)} = 3.84$, $0.0001 < p < 0.001$, reject H_0
Potato salad, $x^2 = 8.489$, $x^2_{1,(.05)} = 3.84$, $0.001 < p < 0.01$, reject H_0
Cole slaw, $x^2 = 1.482$, $x^2_{1,(.05)} = 3.84$, $0.1 < p < 0.5$, fail to reject H_0

7.19 $x^2 = 3.887$, $x^2_{2,(.05)} = 5.99$, $0.1 < p < 0.5$, fail to reject H_0

7.21 $x^2 = 32.04$, $x^2_{3,(.05)} = 7.81$, $p < 0.0001$, reject H_0

7.23 $x^2 = 18.529$, $x^2_{2,(.05)} = 5.99$, $p < 0.0001$, reject H_0

7.25

	Polio	Without Polio	Total
Vaccine	57	200688	200745
	99.38	200645.62	
	18.073	0.009	
Placebo	142	201087	201229
	99.62	201129.38	
	18.029	0.009	
Total	199	401775	401974

Chi-Sq = 36.120, DF = 1, P-Value < 0.001

Chapter 8

8.1 a) 0.0764, b) 0.9236, c) 0.4096, d) 0.8593

8.3 a) P (Z > 1.04) = 0.1492

b) standard error of the mean = 1.73609

P (Z > 3.29) = 0.0005

c) $P(-0.42 < Z < 1.40) = 0.5820$

8.5 a) P (Z > 0.05) = 0.4801

b) 95% CL = (2670g to 4482g)

8.7

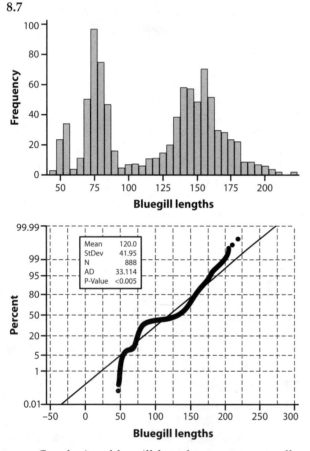

Conclusion: bluegill lengths are not normally distributed

8.9 a) $P(x \le 8) = 0.71928$

b) $P(z < 0.34) = 0.1331 + 0.5 = 0.6331$

8.11 Using Eq. 8.10: $0.0738 < p < 0.1263$

Using Table A.11: $0.0757 < p < 0.1297$

8.13 $H_0: p_s = p_{ns}$ vs. $H_a: p_s > p_{ns}$

Z = 1.51 (CV = 1.96); Fail to reject H_0

Chapter 9

9.1 95% CI = 66.28, 74.12

9.3 95% CI = 17.28, 21.23

99% CI = 16.39, 22.12

9.5 $H_0: \mu = 100$ vs. $H_a: \mu \ne 100$

$|t| = 5.367$, t critical = 2.093, $p < 0.0001$, reject H_0; mean weight differs from 100; so this hen should be removed from breeding program

9.7 $H_0: \mu \le 3.5$ vs. $H_a: \mu > 3.5$

$t = 1.5811$, t critical = 1.833, $0.05 < p < 0.10$, fail to reject H_0, no evidence of any difference from claim

9.9 $H_0: \mu \ge 50$ vs. $H_a: \mu < 50$

$t = -2.8946$; t critical = -1.729; $0.0005 < p < 0.005$; reject H_0; enzyme levels are depressed, suggesting that water is polluted

9.11 $H_0: \mu \ge 8.0$ vs. $H_a: \mu < 8.0$

$t = -1.3176$, t critical = -1.729, $p > 0.1$, fail to reject H_0, no evidence that the mean weight is less than 8.0 kg

9.13 $H_0: \mu(\text{post} - \text{pre}) = 0$ vs. $H_a: \mu(\text{post} - \text{pre}) \ne 0$

$t = 4.2668$, t critical = 2.365, $0.001 < p < 0.01$, reject H_0, post exercise pulse rate differs from pre-exercise pulse rate

9.15 $H_0: \mu(\text{cort} - \text{plac}) \ge 0$ vs. $H_a: \mu(\text{cort} - \text{plac}) < 0$

$t = -5.745$, t critical = -1.796, $0.00005 < p < 0.0005$, reject H_0, cortisone decreases the retention rate of words

9.17

Dif (After-Before)	Rank
4	3
41	9
29	6
12	4
40	8
35	7
−1	1.5 (−)
68	11
69	12
−1	1.5 (−)
59	10
27	5

$H_0: \theta_{\text{dif}} \le 0$ vs. $H_a: \theta_{\text{dif}} > 0$

T = 3, t critical = 17; $p < 0.005$, reject H_0, after receiving the flu vaccine, individuals increase the rate of encounters

9.19 H_0: The drug has no effect on one's sense of well-being vs.

H_a: The drug has an effect on one's sense of well-being

$k = 8$, $p = 0.5$, $P(X \le 1) = 0.035156$ (p-value, one-tailed), reject H_0

Chapter 10

10.1 H_0: μ(Bog A) = μ(Bog B) vs. H_a: μ(Bog A) ≠ μ(Bog B)
$|t|$= 0.555, $p > 0.2$, t critical = 2.064, fail to reject H_0, mean weight does not differ between bogs

10.3 a. H_0: μ(men) = μ(women) vs. H_a: μ(men) ≠ μ(women)
$|t|$= 3.833, t critical = 2.003, $0.0001 < p < 0.001$, reject H_0, the mean pulse rate differs between men and women
b. 95% CI for $\mu_w - \mu_m$ = 4.049, 12.913
c. We are 95% sure that the mean resting pulse rate is between 4.0 and 12.9 beats per minute greater in women than in men.

10.5 H_0: μ(down) ≤ μ(up) vs. H_a: μ(down) > μ(up)
t = 5.140, t critical = 1.895, $0.0005 < p < 0.005$, reject H_0, the mean liver alcohol dehydrogenase activity is greater in catfish downstream than catfish upstream

10.7 H_0: μ(glucose) = μ(sucrose) vs. H_a: μ(glucose) ≠ μ(sucrose)
t = 0.731, t critical = 2.77, $p > 0.2$, fail to reject H_0, there is no evidence that growth rate differs between the two carbon sources

10.9 H_0: μ(athletes) ≥ μ(non) vs. H_a: μ(athletes) < μ(non)
t = –2.216, t critical = –1.711, $0.01 < p < 0.025$, reject H_0, male athletes have a faster reaction time than male nonathletes

10.11 H_0: θ(hairy) ≤ θ(nonhairy) vs. H_a: θ(hairy) > θ(nonhairy)
$\sum R_a$ = 146.5, $\sum R_b$ = 63.5, U = 91.5, U_{crit} = 73, $p \ll 0.05$, reject H_0, hairy appear scarier!

10.13 H_0: θ(control) = θ(treated) vs. H_a: θ(control) ≠ θ(treated)
$\sum R_a$ = 100.5, $\sum R_b$ = 202.5, U = 133.5, U_{crit} = 115, $p < 0.05$, reject H_0, PHA affects blastogenesis

10.15 Mann-Whitney U Test
H_0: θ(sprayed) = θ(unsprayed) vs. H_a: θ(sprayed) ≠ θ(unsprayed)
$\sum R_a$ = 94.5, $\sum R_b$ = 181.5, U = 90.5, U_{crit} = 97, $p > 0.05$, fail to reject H_0, not enough evidence to suggest that there is a difference between sprayed and unsprayed areas

10.17 Two-sample t test
H_0: μ(infected) = μ(healthy) vs. H_a: μ(infected) ≠ μ(healthy)
$|t|$ = 3.715, t critical = 2.571, reject H_0, infection by the parasite affects spleen weight in this species
95% CI for $\mu_i - \mu_h$= 0.956, 5.253

Chapter 11

11.1 H_0 all μ's equal (no effect of virus on enzyme activity) vs.
H_a: one or more μ's different (virus affects enzyme activity)
$F_{2, 12, (.05)}$ = 3.89, F_{stat} = 30.87, $p \ll 0.05$, reject H_0

Source	Sum of Squares	df	MS	F
Between-Groups	6.2713	2	3.1356	30.87
Error	1.2191	12	0.1016	
Total	7.4904	14		

Tukey Test:
CV = 0.5374
Control – TMV = 1.034*
Control – TRSV = 1.556*
TMV – TRSV = 0.522
Infection by both viruses affects mean enzyme activity.

11.3 H_0: no variation in cellulase activity among fungus strains
H_a: variation in cellulase activity among fungus strains
$F_{3,16, (.05)}$ ≈ 3.29, F_{stat} = 70.353, $p \ll 0.01$, reject H_0

Source	Sum of Squares	df	MS	F
Between-Groups	632.95	3	210.98	70.33
Error	48.00	16	3.00	
Total	680.95	19		

Conclusion: cellulose activity is variable among fungus strains.

11.5 H_0: all θ's equal (no difference in aggressiveness among conditions) vs.
H_a: one or more θ's different (aggressiveness differs among conditions)
H = 10.81, $\chi^2_{2, .05}$ = 5.99, $0.001 < p < 0.01$, reject H_0

Conclusion: aggressiveness differs among the three conditions

11.7 H_0: all θ's equal (brain size unaffected by benzene) vs.
H_a: one or more θ's different (brain size is affected by benzene)
H = 10.385, $\chi^2_{2, .05}$ = 5.99, $0.001 < p < 0.01$, reject H_0

Conclusion: brain size in these frogs is affected by benzene during development

11.9 Checking Assumptions:
Normality Test: Anderson-Darling Test (p = 0.983)—passes
Homogeneous Variance: Barlett's Test (p = 0.674), Levene's Test (p = 0.617)—passes

H_0: No added variation in selenium content in zooplankton from different lakes
H_a: Zooplankton selenium content varies among lakes

Source	DF	SS	MS	F	P
Between-Groups	4	3785.7	946.4	72.78	0.000
Error	35	455.1	13.0		
Total	39	4240.8			

$p < 0.001$, reject H_0

Conclusion: zooplankton selenium content varies among lakes

11.11 Normality Test: Anderson-Darling Test ($p = 0.532$)

Test for Homogeneous Variance: Barlett's Test ($p = 0.023$), Levene's Test ($p = 0.108$) fails test for homogeneous variance

H_0: median plaque density does not differ among treatments
H_a: treatment affects median plaque density

Kruskal-Wallis Test

	N	Median	Ave Rank	Z
Control	10	370.00	24.6	4.00
Epinephrin	10	117.50	13.7	-0.79
Norepinephrin	10	82.50	8.2	-3.21
Overall	30		15.5	

H = 17.98 DF = 2 P = 0.000

$p < 0.001$, reject H_0

Conclusion: treatment affects median plaque density

11.13 Normality Test: Anderson-Darling Test ($p < 0.005$)

Test for Homogeneous Variance: Barlett's Test ($p = 0.225$), Levene's Test ($p = 0.446$) *fails the normality test so log transformation performed

After Log Transformation:
Normality Test: Anderson-Darling Test ($p = 0.307$)

H_0 all μ's equal (mean antibody response is the same for 3 groups) vs.
H_a: one or more μ's different (mean antibody response differs among treatments)

ANOVA run on the transformed data:

Source	DF	SS	MS	F	P
Between-Groups	2	0.519	0.259	1.66	0.199
Error	54	8.413	0.156		
Total	56	8.932			

$p = 0.199$, fail to reject H_0

Conclusion: mean antibody response is the same for 3 groups of mice

Chapter 12

12.1 1) H_0: no difference among mercury treatments vs. H_a: there is a difference among mercury treatments
2) H_0: no variation among litters vs. H_a: variation among litters

Analysis of Variance for Activity Level

Source	DF	SS	MS	F	P
Treatment	2	3169.4	1584.7	15.63	0.001
Litter	5	15573.6	3114.7	30.72	0.000
Error	10	1013.9	101.4		
Total	17	19756.9			

Based on the low p-values, we can reject both null hypotheses.

Conclusion: Both mercury and litter affect activity levels, with the highest activity levels observed in mice exposed to Mercuric Chloride.

12.3 Randomized Block Design Using Repeated Measures
1) H_0: no effect of temperature treatment vs. H_a: effect of temperature treatment
2) H_0: no variation among plants vs. H_a: variation among plants

Analysis of Variance for Rate of Photosynthesis

Source	DF	SS	MS	F	P
Block (Plant)	4	39876.3	9969.1	270.41	0.000
Treatment	2	1044.4	522.2	14.16	0.002
Error	8	294.9	36.9		
Total	14	41215.6			

Based on the low p-values, we can reject both null hypotheses.

Conclusion: Both temperature treatment and plant affect photosynthetic rate, with photosynthetic rate decreasing one day after treatment but then returning to pretreatment levels one week after treatment.

12.5 1) H_0: no difference among depths vs. H_a: there is a difference among depths
2) H_0: no variation among lakes vs. H_a: variation among lakes
$F_{3,6, (0.05)} = 4.76$, reject H_0

Analysis of Variance for Algae Concentration

Source	DF	SS	MS	F	P
Lake	3	84247	28082	5.56	0.036
Depth	2	178189	89094	17.63	0.003
Error	6	30322	5054		
Total	11	292757			

Based on the low p-values, we can reject both null hypotheses.

Conclusion: Both depth and lake affect algae concentration, with the surface having the greatest concentration of algae.

12.7 H_0: no difference over time vs. H_a: there is a difference over time
$x^2_{4, (.05)} = 9.49$

```
S = 22.10  DF = 4  P = 0.000
S = 22.29  DF = 4  P = 0.000 (adjusted for ties)

Est       Sum of
Interval    N   Median   Ranks
1           6      18    29.0
2           6       7    24.0
3           6       5    18.5
4           6       3    12.5
5           6       1     6.0

Grand median = 7
```

Reject H_0.

Conclusion: The interval number significantly affects the number of copulatory attempts, with the number of copulatory attempts decreasing with time.

12.9 H_0: no difference between sex/reproductive groups vs. H_a: there is a difference between sex/reproductive groups ($F_{2,45, (.05)} = 3.21$)
H_0: no effect of crowding vs. H_a: there is an effect of crowding ($F_{2,45, (.05)} = 3.21$)
H_0: no interaction between sex and crowding vs. H_a: there is an interaction effect ($F_{4,45, (.05)} = 2.59$)

```
Analysis of Variance for Corticosterone

Source       DF    SS      MS      F       P
Sex           2  16864    8432   51.35  0.000
Crowding      2 473126  236563 1440.73  0.000
Sex*Crowding  4   4401    1100    6.70  0.000
Error        45   7389     164
Total        53 501780
```

Based on the low p-values, we can reject all three null hypotheses.

Conclusion: The level of crowding and the sex/reproductive status significantly affect the plasma corticosterone levels, with individuals experiencing more crowding having higher levels of plasma corticosterone and the effect is greatest on gravid females. The strong interaction effect indicates that the response to crowding depends on sex and the differences between sex/reproductive groups depends on crowding levels.

12.11 H_0: no effect of age vs. H_a: there is an effect of age ($F_{2,24, (.05)} = 3.40$)
H_0: no effect of sex vs. H_a: there is an effect of sex ($F_{1,24, (.05)} = 4.26$)

H_0: no interaction between age and sex vs. H_a: there is an interaction effect ($F_{2,24, (.05)} = 3.40$)

```
Analysis of Variance for Blood Pressure
Source   DF    SS      MS     F      P
Age       2  6661.7  3330.8  46.90  0.000
Sex       1   616.5   616.5   8.68  0.007
Age*Sex   2   622.1   311.0   4.38  0.024
Error    24  1704.4    71.0
Total    29  9604.7
```

Since all three p-values are small, we can reject all three null hypotheses.

Conclusion: The age and sex of hamsters significantly affect the systolic blood pressure, with males having a higher systolic blood pressure than females when they are mature and old, but males have a lower systolic blood pressure than females when they are adolescents.

12.13 Randomized Complete Block Design with Repeated Measures
H_0: no effect of caffeine on pulse rate vs. H_a: caffeine affects pulse rate ($F_{2,10, (0.05)} = 4.10$)
H_0: no variation among rats vs. H_a: variation among rats ($F_{5,10, (0.05)} = 3.33$)

```
Analysis of Variance for Pulse rate

Source      DF    SS      MS      F      P
Rat(block)   5 1369.11 273.82 125.10  0.000
Treatment    2  483.44 241.72 110.43  0.000
Error       10   21.89   2.19
Total       17 1874.44
```

Reject both null hypotheses.

Conclusion: Caffeine treatment affects pulse rate, with an observed increase in pulse rate after being administered caffeine. Furthermore, individual rats vary in their pulse rates.

12.15 Friedman Test
H_0: no preference for a particular prey vs. H_a: there is a preference for a particular prey
$x^2_{3, (.05)} = 7.81$

Friedman Test: Number Eaten versus Prey blocked by Naiad

```
S = 11.85  DF = 3  P = 0.008
S = 12.05  DF = 3  P = 0.007 (adjusted for ties)

Est      Sum of
Prey       N    Median   Ranks
Prey A     6     8.313    23.0
Prey B     6     3.563    17.0
Prey C     6     1.688     9.5
Prey D     6     1.688    10.5

Grand median = 3.813
```

Conclusion: damselflies show preference according to prey type. Based on the sample medians, they appear to prefer prey type A.

12.17 Randomized Complete Block Design with Repeated Measures

H_0: no difference in testosterone levels among seasons vs. H_a: there is a difference among season

$(F_{3,15, (0.05)} = 3.29)$

H_0: no variation among hogs in testosterone levels vs. H_a: variation among hogs

$(F_{5,15, (0.05)} = 2.90)$

```
Analysis of Variance for Testosterone
Source   DF      SS      MS      F       P
Male     5     2461     492   0.43   0.820
Season   3    95034   31678  27.71   0.000
Error   15    17149    1143
Total   23   114644
```

Conclusion: (1) Season does have an effect on testosterone levels, with testosterone levels being greatest in the fall. (2) No significant variation among males in their testosterone levels.

Chapter 13

13.1 H_0: $\rho = 0$ vs. H_a: $\rho \neq 0$

$r = 0.747$, $p < 0.001$, reject H_0, systolic and diastolic blood pressure are correlated

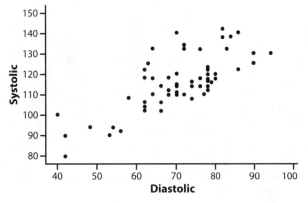

13.3 H_0: $\rho = 0$ vs. H_a: $\rho \neq 0$

$r = 0.994$, $p < 0.001$, reject H_0, method one and method two are correlated

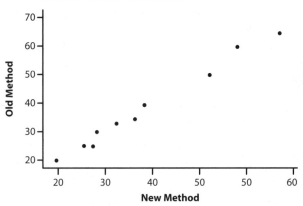

13.5 H_0: $\rho = 0$ vs. H_a: $\rho \neq 0$

body vs. spleen: $r = -0.049$, $p = 0.900$, fail to reject H_0, no correlation between spleen weight and body weight.

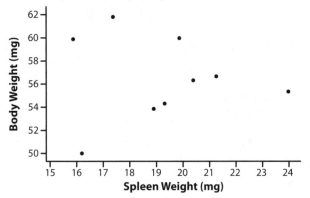

body vs. bursa: $r = 0.141$, $p = 0.717$, fail to reject H_0, no correlation

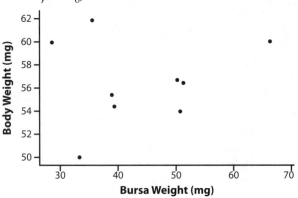

bursa vs. spleen: $r = 0.450$, $p = 0.225$, fail to reject H_0, no correlation

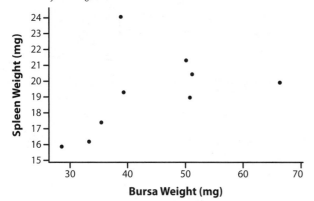

13.7 $\sum d^2 = 321.5$, $r_s = 0.5272$, $t_{stat} = 2.3215$, t critical $= 2.145$, $0.02 < p < 0.05$. The correlation is significant, indicating a positive association between change in activity and body size. Inspection of the graph indicates that the direction of the effect is as expected.

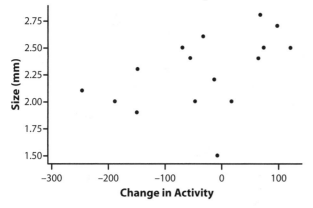

Chapter 14

14.1 $H_0: \beta = 0$ vs. $H_a: \beta \neq 0$

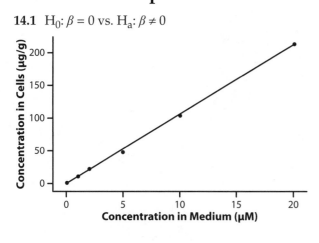

Source	DF	SS	MS	F	P
Regression	1	33093	33093	4750.44	0.000
Residual Error	4	28	7		
Total	5	33121			

$F_{1,4,(.05)} = 7.71$; $p < 0.001$, reject H_0. Conclusion: concentration in cells clearly depends on concentration in medium.

$\hat{y} = -1.90 + 10.7x$; $r^2 = 0.999$

14.3 $H_0: \beta = 0$ vs. $H_a: \beta \neq 0$

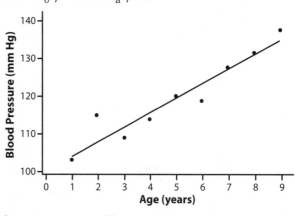

Source	DF	SS	MS	F	P
Regression	1	912.60	912.60	70.23	0.000
Residual Error	7	90.96	12.99		
Total	8	1003.56			

$F_{1,7,(.05)} = 5.59$; $p < 0.001$, reject H_0. Conclusion: blood pressure clear increases with age of road warblers.

$\hat{y} = 100 + 3.90x$; $r^2 = 0.909$

x	$\hat{Y}$	±	95% CI for $\hat{y}$	
1	103.900	1.80948	102.091	105.709
2	107.800	1.34419	106.456	109.144
3	111.700	0.87889	110.821	112.579
4	115.600	0.41360	115.186	116.014
5	119.500	0.05170	119.448	119.552
6	123.400	0.51700	122.883	123.917
7	127.300	0.98229	126.318	128.282
8	131.200	1.44759	129.752	132.648
9	135.100	1.91288	133.187	137.013

14.5 $H_0: \beta = 0$ vs. $H_a: \beta \neq 0$

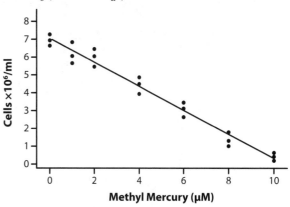

```
Source          DF    SS     MS     F      P
Regression       1 114.32 114.32 698.74 0.000
Residual Error  19   3.11   0.16
Total           20 117.43
```

$F_{1,19,(.05)} = 4.38$; $p < 0.001$, reject H_0. Conclusion: cell density declines with mercury concentration.

$\hat{y} = 6.99 - 0.675x$; $r^2 = 0.974$

14.7 H_0: $\beta = 0$ vs. H_a: $\beta \neq 0$

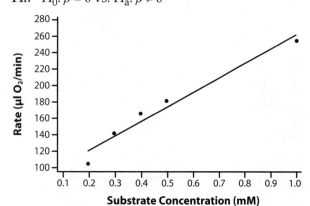

Let $y = 1/\text{rate}$, $x = 1/\text{concentration}$

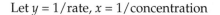

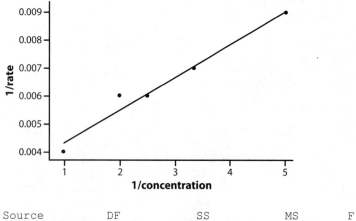

```
Source          DF        SS              MS          F      P
Regression       1  0.000012833    0.000012833  104.99  0.002
Residual Error   3  0.000000367    0.000000122
Total            4  0.000013200
```

$F_{1,3,(.05)} = 10.1$; $p = 0.002$, reject H_0. Conclusion: reaction rate depends on substrate concentration.

$\hat{y} = 0.00311 + 0.00119x$; $r^2 = 0.972$

References on Statistics, Experimental Design, and Applications

Ambrose, H. W., and K. P. Ambrose. *A Handbook of Biological Investigation*, 3rd ed. Winston-Salem, NC: Hunter Textbooks, Inc., 1981.

Box, G. E. P., W. G. Hunter, and J. S. Hunter. *Statistics for Experimenters: An Introduction to Design, Data Analysis, and Model Building*. Hoboken, NJ: John Wiley & Sons, Inc., 1978.

Forthofer, R. N., and E. S. Lee. *Introduction to Biostatistics: A Guide to Design, Analysis, and Discovery*. San Diego: Academic Press, 1995.

Gonick, L., and W. Smith. *The Cartoon Guide to Statistics*. New York: Harper Perennial, 1993.

Good, P. I. *Resampling Methods: A Practical Guide to Data Analysis*, 3rd ed. New York: Birkhäuser Boston, 2006.

Gotelli, N. J., and A. M. Ellison. *A Primer of Ecological Statistics*. Sunderland, MA: Sinauer Associates, Inc., 2004.

Green, R. H. *Sampling Design and Statistical Methods for Environmental Biologists*. Hoboken, NJ: John Wiley & Sons, Inc., 1979.

Hamilton, J. D. *Time Series Analysis*. Princeton, NJ: Princeton University Press, 1994.

Hollander, M., and D. A. Wolfe. *Nonparametric Statistical Methods*. Hoboken, NJ: John Wiley & Sons, Inc., 1973.

Lilienfeld, A. M., and D. E. Lilienfeld. *Foundations of Epidemiology*, 2nd ed. New York: Oxford University Press, 1980.

McGarigal, K., S. Cushman, and S. Stafford. *Multivariate Statistics for Wildlife and Ecology Research*. New York: Springer-Verlag New York, Inc., 2000.

Mead, R. *The Design of Experiments: Statistical Principles for Practical Application*. Cambridge: Cambridge University Press, 1980.

Neter, J., W. Wasserman, and M. H. Kutner. *Applied Linear Regression Models*, 2nd ed. Scarborough, ON: Irwin, 1989.

Quinn, G. P., and M. J. Keough. *Experimental Design and Data Analysis for Biologists*. Cambridge: Cambridge University Press, 2002.

Rohlf, F. J., and R. R. Sokal. *Statistical Tables*, 4th ed. New York, NY: W.H. Freeman, 2012.

Scheiner, S. M., and J. Gurevitch, eds. *Design and Analysis of Ecological Experiments*. New York: Chapman and Hall, 1993.

Siegel, S. *Nonparametric Statistics for the Behavioral Sciences*. New York: McGraw-Hill, 1956.

Siegel, S., and N. J. Castellan. *Nonparametric Statistics for the Behavioral Sciences*, 2nd ed. New York: McGraw-Hill, 1988.

Sokal, R. R., and F. J. Rohlf. *Biometry*, 4th ed. New York: W.H. Freeman, 2012.

Snedecor, G. W., and W. G. Cochran. *Statistical Methods*, 8th ed. Ames: Iowa State University Press, 1989.

Stevens, J. *Applied Multivariate Statistics for the Social Sciences*, 3rd ed. Mahwah, NJ: Lawrence Erlbaum Associates, 1996.

Whitlock, M. C., and D. Schluter. *The Analysis of Biological Data*. Greenwood Village, CO: Roberts & Company Publishers, 2008.

Wonnacott, T. H., and R. J. Wonnacott. *Introductory Statistics for Business and Economics*, 2nd ed. Hoboken, NJ: John Wiley & Sons, Inc., 1977.

Zar, J. H. 1999. *Biostatistical Analysis*, 4th ed. Upper Saddle River, NJ: Prentice-Hall, 1999.

The Biggest Public Health Experiment Ever
The 1954 Field Trial of the Salk Poliomyelitis Vaccine[1]

Paul Meier
University of Chicago

The largest and, until the 1980s, the most expensive medical experiment in history was carried out in 1954. Well over a million young children participated, and the immediate direct costs were over 5 million midcentury dollars. The experiment was carried out to assess the effectiveness, if any, of the Salk vaccine as a protection against paralysis or death from poliomyelitis. The study was elaborate in many respects, most prominently in the use of placebo controls (children who were inoculated with simple salt solution) assigned at random (that is, by a carefully applied chance process that gave each volunteer an equal probability of getting vaccine or salt solution) and subjected to a double-blind evaluation (that is, an arrangement under which neither the children nor the physicians who evaluated their subsequent state of health knew who had been given the vaccine and who the salt solution).

Why was such elaboration necessary? Did it really result in more or better knowledge than could have been obtained from much simpler studies? These are the questions on which this discussion is focused.

Background

Polio was never a common disease, but it certainly was one of the most frightening and, in many ways, one of the most inexplicable in its behavior. It struck hardest at young children, and, although it was responsible for only about 6% of the deaths in the age group 5 to 9 in the early 1950s, it left many helpless cripples, including some who could survive only in a respirator. It appeared in epidemic waves, leading to summer seasons in which some communities felt compelled to close swimming pools and restrict public gatherings as cases increased markedly from week to week; other communities, escaping an epidemic one year, waited in trepidation for the year in which their turn would come. Rightly or not, this combination of selective attack upon the most helpless age group, and the inexplicable vagaries of its epidemic behavior, led to far greater concern about polio as a cause of death than other causes, such as auto accidents, which are more frequent and, in some ways, more amenable to community control.

The determination to mount a major research effort to eradicate polio arose in no small part from the involvement of President Franklin D. Roosevelt, who was struck down by polio when a successful young politician. His determination to overcome his paralytic handicap and the commitment to the fight against polio made by Basil O'Connor, his former law partner, enabled a great deal of attention, effort, and money to be expended on the care and rehabilitation of polio victims and—in the end, more importantly—on research into the causes and prevention of the disease.

During the course of this research, it was discovered that polio is caused by a virus and that three main virus types are involved. Although clinical manifestations of polio are rare, it was discovered that the virus itself was not rare, but common, and that most adults had experienced a polio infection sometime in their lives without ever being aware of it.

This finding helped to explain the otherwise peculiar circumstance that polio epidemics seemed to hit hardest those who were better off hygienically (that is, those who had the best nutrition, most favorable housing conditions, and were otherwise apparently most favorably situated). Indeed, the disease seemed to be virtually unknown in those countries with the poorest hygiene. The explanation is that because there was plenty of polio virus in the less-favored populations, almost every infant was exposed to the disease early in life while still protected by the immunity passed on from the mother. As a result, everyone had had polio, but under protected circumstances, and, thereby, everyone had developed immunity.

As with many other virus diseases, an individual who has been infected by polio and recovered is usually immune to another attack (at least by a virus strain of the same type). The reason for this is that the body, in fighting the infection, develops *antibodies* (a part of the gamma globulin fraction of the blood) to the *antigen* (the protein part of the polio virus). These antibodies remain in the bloodstream for years, and even when their level declines so far as to be scarcely measurable, there are usually enough of them to prevent a serious attack from the same virus.

Smallpox and influenza illustrate two different approaches to the preparation of an effective vaccine. For smallpox, which has long been controlled by a vaccine, we use for the vaccine a closely related virus, cowpox, which is ordinarily incapable of causing serious disease in humans, but which gives rise to antibodies that also protect against smallpox. (In a very few individuals this vaccine is capable of causing a severe, and occasionally fatal, reaction. The risk is small enough, however, so that before smallpox was conquered we did not hesitate to expose all our schoolchildren to it in order to protect them from smallpox.) In the case of influenza, however, instead of a closely related live virus, the vaccine is a solution of the influenza virus itself, prepared with a virus that has been killed by treatment with formaldehyde. Provided that the treatment is not too prolonged, the dead virus still has enough antigenic activity to produce the required antibodies so that, although it can no longer infect, it is sufficiently like the live virus to be a satisfactory vaccine.

For polio, both of these methods were explored. A live-virus vaccine would have the advantage of reproducing in the vaccinated individual and, hopefully, giving rise to a strong reaction that would produce a high level of long-lasting antibodies. With such a vaccine, however, there might be a risk that a vaccine virus so similar to the virulent polio virus could mutate into a virulent form and itself be the cause of paralytic or fatal disease. A killed-virus vaccine should be safe because it presumably could not infect, but it might fail to give rise to an adequate antibody response. These and other problems stood in the way of the rapid development of a successful vaccine. Some unfortunate prior experience also contributed to the cautious approach of the researchers. In the 1930s, attempts

had been made to develop vaccines against polio; two of these were actually in use for a time. Evidence that at least one of these vaccines had been responsible for cases of paralytic polio soon caused both to be promptly withdrawn from use. This experience was very much in the minds of polio researchers, and they had no wish to risk a repetition.

Research to develop both live and killed vaccines was stimulated in the late 1940s by the development of a tissue culture technique for growing polio virus. Those working with live preparations developed harmless strains from virulent ones by growing them for many generations in suitable tissue culture media. There was, of course, considerable worry lest these strains, when used as a vaccine in humans, might revert to virulence and cause paralysis or death. (It's now clear that the strains developed are indeed safe—a live-virus preparation taken orally is the vaccine presently in widespread use throughout the world.)

Those working with killed preparations, notably Jonas Salk, had the problem of treating the virus (with formaldehyde) sufficiently to eliminate its infectiousness, but not so long as to destroy its antigenic effect. This was more difficult than expected, and some early lots of the vaccine proved to contain live virus capable of causing paralysis and death. There are statistical issues in the safety story (Meier, 1957), but our concern here is with the evaluation of effectiveness.

Evaluation of Effectiveness

In the early 1950s the Advisory Committee convened by the National Foundation for Infantile Paralysis (NFIP) decided that the killed-virus vaccine developed by Jonas Salk at the University of Pittsburgh had been shown to be both safe and capable of inducing high levels of the antibody in children on whom it had been tested. This made the vaccine a promising candidate for general use, but it remained to prove that the vaccine actually would prevent polio in exposed individuals. It would be unjustified to release such a vaccine for general use without convincing proof of its effectiveness, so it was determined that a large-scale "field trial" should be undertaken.

That the trial had to be carried out on a very large scale is clear. For suppose we wanted the trial to be convincing if indeed the vaccine were 50% effective (for various reasons, 100% effectiveness could not be expected). Assume that, during the trial, the rate of occurrence of polio would be about 50 per 100,000 (which was about the average incidence in the United States during the 1950s). With 40,000 in the control group and 40,000 in the vaccinated group, we would find about 20 control cases and about 10 vaccinated cases, and a difference of this magnitude could fairly easily be attributed to random variation. It would suggest that the vaccine might be effective, but it would not be persuasive. With 100,000 in each group, the expected numbers of polio cases would be 50 and 25, and such a result would be persuasive. In practice, a much larger study was clearly required because it was important to get definitive results as soon as possible, and if there were relatively few cases of polio in the test area, the expected number of cases might be well under 50. It seemed likely, also, for reasons we shall discuss later, that paralytic polio, rather than all polio, would be a better criterion of disease, and only about half the diagnosed cases are classified "paralytic." Thus the relatively low incidence of the disease, and its great variability from place to place and time to time, required that the trial involve a huge number of subjects—as it turned out, over a million.

The Vital Statistics Approach

Many modern therapies and vaccines, including some of the most effective ones such as small-pox vaccine, were introduced because preliminary studies suggested their value. Large-scale use subsequently provided clear evidence of efficacy. A natural and simple approach to the evaluation of the Salk vaccine would have been to distribute it as widely as possible, through the schools, to see whether the rate of reported polio was appreciably less than usual during the subsequent season. Alternatively, distribution might be limited to one or a few areas because limitations of supply would preclude effective coverage of the entire country. There is even a fairly good chance that were one to try out an effective vaccine against the common cold, convincing evidence might be obtained in this way.

In the case of polio—and, indeed, in most cases—so simple an approach would almost surely fail to produce clear-cut evidence. First, and foremost, we must consider how much polio incidence varies from season to season, even without any attempts to modify it. From figure 1, which shows the annual reported incidence from 1930 through 1955, we see that had a trial been conducted in this way in 1931, the drop in incidence from 1931 to 1932 would have been strongly suggestive of a highly effective vaccine because the incidence dropped to

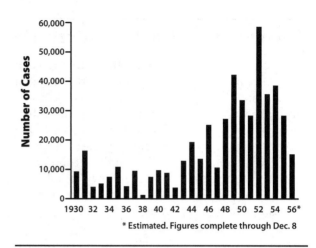

Figure 1 Poliomyelitis in the U.S., 1930–56. Source: Rutstein (1957)

less than a third of its previous level. Similar misinterpretations would have been made in 1935, 1937, and other years—for example, 1952. One might suppose that such mistakes could be avoided by using the vaccine in one area, say, New York State, and comparing the rate of incidence there with that of an unvaccinated area, say, Illinois. Unfortunately, an epidemic of polio might well occur in Chicago—as it did in 1956—during a season in which New York had a very low incidence.

Another problem, more subtle, but equally burdensome, relates to the vagaries of diagnosis and reporting. There is no difficulty, of course, in diagnosing the classic respirator case of polio, but the overwhelming majority of cases are less clear-cut. Fever and weakness are common symptoms of many illnesses, including polio, and the distinction between weakness and slight transitory paralysis will be made differently by different observers. Thus the decision to diagnose a case as nonparalytic polio instead of some other disease might well be influenced by a physician's general knowledge or feeling about how widespread polio is in his or her community at the time.

These difficulties can be mitigated to some extent by setting down very precise criteria for diagnosis, but it is virtually impossible to obviate them completely when, as would be the case after the widespread introduction of a new vaccine, there is a marked shift in what the physician expects to find. This is most especially true when the initial diagnosis must be made by family physicians who cannot easily be indoctrinated in the use of a special set of criteria, as is the case with polio. Later evaluation by specialists cannot, of course, bring into the picture those cases originally diagnosed as something other than polio.

The Observed Control Approach

The difficulties of the vital statistics approach were recognized by all concerned, and the initial study plan, although not judged entirely satisfactory, got around many of the problems by introducing a control group similar in characteristics to the vaccinated group. More specifically, the idea was to offer vaccination to all children in the second grade of participating schools and to follow the polio experience not only in these children but in the first- and third-grade children as well. Thus the vaccinated second-graders would constitute the *treated* group, and the first- and third-graders would constitute the *control* group. This plan follows what we call the *observed control approach.*

It is clear that this plan avoids many of the difficulties listed above. The three grades all would be drawn from the same geographic location so that an epidemic affecting the second grade in a given school would certainly affect the first and third grades as well. Of course, all subjects would be observed concurrently in time. The grades, naturally, would be different ages, and polio incidence does vary with age. Not much variation from grade to grade was expected, however, so it seemed reasonable to assume that the average of first and third grades would provide a good control for the second grade.

Despite the relative attractiveness of this plan and its acceptance by the NFIP advisory committee, serious objections were raised by certain health departments that were expected to participate. In their judgment, the results of such a study were likely to be insufficiently convincing for two important reasons. One is the uncertainty in the diagnostic process mentioned earlier and its liability to influence by the physician's expectations, and the other is the selective effect of using volunteers.

Under the proposed study design, physicians in the study areas would have been aware of the fact that only second-graders were offered vaccine, and in making a diagnosis for any such child, they would naturally and properly have inquired whether the child had been vaccinated. Any tendency to decide a difficult diagnosis in favor of nonpolio when the child was known to have been vaccinated would have resulted in a spurious piece of evidence favoring the vaccine. Whether or not such an effect was really operating would have been almost impossible to judge with assurance, and the results, if favorable, would have been forever clouded by uncertainty.

A less conjectural difficulty lies in the difference between those families who volunteer their children for participation in such a trial and those who do not. Not at all surprisingly, it was later found that those who do volunteer tend to be better educated and, generally, more well-to-do than those who do not participate. There was also evidence that those who agree to participate tend to be absent from school with a noticeably higher frequency than others. The direction of effect of such selection on the incidence of diagnosed polio is by no means clear before the fact, and this important difference between the treated group and the control group also would have clouded the interpretation of the results.

Randomization and the Placebo Control Approach

The position of critics of the NFIP plan was that the issue of vaccine effectiveness was far too important to be studied in a manner that would leave uncertainties in the minds of reasonable observers. No doubt, if the vaccine should appear to have fairly high effectiveness, most public health officials and the general public would accept it, despite the reservations. If, however, the observed control scheme were used, a number of qualified public health scientists would have remained unconvinced, and the value of the vaccine would be uncertain. Therefore, the critics proposed that the study be run as a scientific experiment with the use of appropriate randomizing procedures to assign subjects to treatment or to control and with a maximum effort to eliminate observer bias. This plan follows what we call the *placebo control approach.*

The chief objection to this plan was that parents of schoolchildren could not reasonably be expected to permit their children to participate in an experiment in which they might be getting only an ineffective salt solution instead of a probably helpful vaccine. It was argued further that the injection of placebo might not be ethically sound since a placebo injection carries a small risk, especially if the child unknowingly is already infected with polio.

The proponents of the placebo control approach maintained that, if properly approached, parents *would* consent to their children's participation in such an experiment, and they judged that because the injections would not be given during the polio season, the risk associated with the placebo injection itself was vanishingly small. Certain health departments took a firm stand: they would participate in the trial only if it were such a well-designed experiment. The consequence was that in approximately half the areas, the randomized placebo control method was used, and in the remaining areas, the alternating-grade observed control method was used.

A major effort was put forth to eliminate any possibility of the placebo control results being contaminated by subtle observer biases. The only firm way to accomplish this was to ensure that neither the subject, nor the parents, nor the diagnostic personnel could know which children had gotten the vaccine until all diagnostic decisions had been made. The method for achieving this result was to prepare placebo material that looked just like the vaccine but was without any antigenic activity, so that the controls might be inoculated and otherwise treated in just the same fashion as were the vaccinated.

Each vial of injection fluid was identified only by a code number so that no one involved in the vaccination or the diagnostic evaluation process could know which children had gotten the vaccine. Because no one knew, no one could be influenced to diagnose differently for vaccinated cases and for controls. An experiment in which both the subject getting the treatment and the diagnosticians who will evaluate the outcome are kept in ignorance of the treatment given each individual is called a *double-blind* experiment. Experience in clinical research has shown the double-blind experiment to be the only satisfactory way to avoid potentially serious observer bias when the final evaluation is in part a matter of judgment.

For most of us, it is something of a shock to be told that competent and dedicated physicians must be kept in ignorance lest their judgments by be colored by knowledge of treatment status. We should keep in mind that it is not deliberate distortion of findings by the physician that concern the medical experimenter. It is rather the extreme difficulty in many cases of making an uncertain decision that, experience has shown, leads the best of investigators to be subtly influenced by information of this kind. For example, in the study of drugs used to relieve postoperative pain, it has been found that it is quite impossible to get an unbiased judgment of the quality of pain relief, even from highly qualified investigators, unless the judge is kept in ignorance of which patients were given the drugs.

The second major feature of the experimental method was the assignment of subjects to treatments by a careful randomization procedure. As we observed earlier, the chance of coming down with a diagnosed case of polio varies with a great many factors including age, socioeconomic status, and the like. If we were to make a deliberate effort to match up the treatment and control groups as closely as possible, we should have to take care to balance these and many other factors, and, even so, we might miss some important ones. Therefore, per-

haps surprisingly, we leave the balancing to a carefully applied equivalent of coin tossing: we arrange that each individual has an equal chance of getting vaccine or placebo, but we eliminate our own judgment entirely from the individual decision and leave the matter to chance.

The gain from doing this is twofold. First, a chance mechanism usually will do a good job of evening out all the variables—those we didn't recognize in advance as well as those we did recognize. Second, if we use a chance mechanism in assigning treatments, we may be confident about the use of the theory of chance (that is, probability theory) to judge the results. We can then calculate the probability that so large a difference as that observed could reasonably be due solely to the way in which subjects were assigned to treatments, or whether, on the contrary, it is really an effect due to a true difference in treatments.

To be sure, there are situations in which a skilled experimenter can balance the groups more effectively than a random-selection procedure typically would. When some factors may have a large effect on the outcome of an experiment, it may be desirable, or even necessary, to use a more complex experimental design that takes account of these factors. However, if we intend to use probability theory to guide us in our judgment about the results, we can be confident about the accuracy of our conclusions only if we have used randomization at some appropriate level in the experimental design.

The final determinations of diagnosed polio proceeded along the following lines. All cases of polio-like illness reported by local physicians were subjected to special examination, and a report of history, symptoms, and laboratory findings was made. A special diagnostic group then evaluated each case and classified it as nonpolio, doubtful polio, or definite polio. The last group was subdivided into nonparalytic and paralytic, with paralytic divided into nonfatal and fatal polio. Only after this process was complete was the code broken and identification made for each case as to whether vaccine or placebo had been administered.

Results of the Trial

The main results are shown in table l, which shows the size of the study populations, the number of cases classified as polio, and the disease rates; that is, the number of cases per 100,000 population. For example, the second line shows that in the placebo control area there were 428 reported

Table 1 Summary of Study Cases by Diagnostic Class and Vaccination Status (rates per 100,000)

Study Group	Study Population	All Reported Cases		Poliomyelitis Cases Total		Paralytic		Non-paralytic		Fatal polio		Not Polio	
		No.	Rate	No.	Rate	No.	Rate	No.	Rate	No.	Rate	No.	Rate
All areas: Total	1,829,916	1,013	55	863	47	685	37	178	10	15	1	150	8
Placebo control areas: Total	749,236	428	57	358	48	270	36	88	12	4	1	70	9
Vaccinated	200,745	82	41	57	28	33	16	24	12	—	—	25	12
Placebo	201,229	162	81	142	71	115	57	27	13	4	2	20	10
Not inoculated*	338,778	182	54	157	46	121	36	36	11	—	—	25	7
Incomplete vaccinations	8,484	2	24	2	24	1	12	1	12	—	—	—	—
Observed control areas: Total	1,080,680	585	54	505	47	415	38	90	8	11	1	80	7
Vaccinated	221,998	76	34	56	25	38	17	18	8	—	—	20	9
Controls†	725,173	439	61	391	54	330	46	61	8	11	2	48	6
Grade 2 not inoculated	123,605	66	53	54	44	43	35	11	9	—	—	12	10
Incomplete vaccinations	9,904	4	40	4	40	4	40	—	—	—	—	—	—

* Includes 8,577 children who received one or two injections of placebo.
† First- and third-grade total population.

Source: Adapted from T. Francis, Jr. (1955), Tables 2 and 3.

cases, of which 358 were confirmed as polio, and, among these, 270 were classified as paralytic (including 4 that were fatal). The third and fourth rows show corresponding entries for those who were vaccinated and those who received placebo, respectively. Beside each of these numbers is the corresponding rate. Using the simplest measure—all reported cases—the rate in the vaccinated group is seen to be half that in the control group (compare the boxed rates in table 1) for the placebo control areas. This difference is greater than could reasonably be ascribed to chance, according to the appropriate probability calculation. The apparent effectiveness of the vaccine is more marked as we move from reported cases to paralytic cases to fatal cases, but the numbers are small and it would be unwise to make too much of the apparent very high effectiveness in protecting against fatal cases. The main point is that the vaccine was a success; it demonstrated sufficient effectiveness in preventing serious polio to warrant its introduction as a standard public health procedure.

Not surprisingly, the observed control area provided results that were, in general, consistent with those found in the placebo control area. The volunteer effect discussed earlier, however, is clearly evident (note that the rates for those not inoculated differ from the rates for controls in both areas). Were the observed control information alone available, considerable doubt would have remained about the proper interpretation of the results.

Although there had been wide differences of opinion about the necessity or desirability of the placebo control design before, there was great satisfaction with the method after the event. The difference between the two groups, although substantial and definite, was not so large as to preclude doubts had there been no placebo controls. Indeed, there were many surprises in the more detailed data. It was known, for example, that some lots of vaccine had greater antigenic power than did others, and it might be supposed that they should have shown a greater protective effect. This was not the case; lots judged inferior in antigenic potency did just as well as those judged superior. Another surprise was the rather high frequency with which apparently typical cases of paralytic polio were not confirmed by laboratory test. Nonetheless, there were no surprises of a character to cast serious doubt on the main conclusion. The favorable reaction of those most expert in research on polio was expressed soon after the results were reported. By carrying out this kind of study before introducing the vaccine, it was noted, we had facts about the Salk vaccine that we still lack about the typhoid vaccine and about the tuberculosis vaccine, after many decades of use.

Epilogue

It would be pleasant to report an unblemished record of success for the Salk vaccine following so expert and successful an appraisal of its effectiveness, but it is more realistic to recognize that such success is but one step in the continuing development of public health science. The Salk vaccine, although a notable triumph in the battle against disease, was relatively crude and, in many ways, not a wholly satisfactory product and it was soon replaced with better ones.

The report of the field trial was followed by widespread release of the vaccine for general use, and it was discovered very quickly that a few of these lots actually caused serious cases of polio. Distribution of the vaccine was then halted while the process was reevaluated. Distribution was reinitiated a few months later, but the momentum of acceptance had been broken, and the prompt disappearance of polio that researchers hoped for did not come about. Meanwhile, research on a more highly purified killed-virus vaccine and on several live-virus vaccines progressed, and within a few years the Salk vaccine was displaced in the United States, but not in Sweden, by live-virus vaccines.

The long-range historical test of the Salk vaccine, in consequence, has never been carried out. We do not know with certainty whether or not that vaccine could have accomplished the relatively complete elimination of polio that has now been achieved. Nonetheless, this does not diminish the importance of its role in providing the first heartening success in the attack on this disease, a role to which careful and statistically informed experimental design contributed greatly.

Problems

1. Using Figure 1 as an example, explain why a control group is needed in experiments where the effectiveness of a drug or vaccine is to be determined.

2. Explain the need for control groups by criticizing the following statement: "A study on the benefits of vitamin C showed that 90% of the people suffering from a cold who take vitamin C get over their cold within a week."

3. Explain the difference between the observed control approach and the placebo control approach. Which one would you prefer, and why?

4. Why is it important to have a double-blind experiment?

5. If double-blind experiments provide the only satisfactory way to avoid observer bias, why aren't they used all the time?

6. If only volunteers are used in an experiment, instead of individuals, will the results of the experiment be of any value? What can you say about the results?

7. Why did the polio epidemics seem to hit hardest those who were better off hygienically?

8. Why was a *large-scale* field trial needed to get convincing evidence of the Salk vaccine's effectiveness?

9. Refer to Figure 1. In which year did the highest polio incidence occur? The lowest? The largest increase? The smallest increase? Give the approximate values of these incidences and increases.

10. Refer to Figure 1. Comment on the use of *the number of cases*. Can you suggest a different indicator of the spread of poliomyelitis in the United States during 1930–1956? When are the two indicators equivalent? (Hint: Refer to table 1.)

Note

[1] Paul Meier. 1989. "The Biggest Public Health Experiment Ever: The 1954 Field Trial of the Salk Poliomyelitis Vaccine." In J. M. Tanur, F. Mosteller, W. H. Kruskal, E. L. Lehmann, R. F. Link, R. S. Pieters, and G. R. Rising (Eds.), *Statistics: A Guide to the Unknown*. Belmont, CA: Wadsworth, pp. 3–14.

References

K. Alexander Brownlee. 1955. "Statistics of the 1954 Polio Vaccine Trials." *Journal of the American Statistical Association* 50(272): 1005–1013.

Thomas Francis, Jr., et al. 1955. "An Evaluation of the 1954 Poliomyelitis Vaccine—Summary Report." *American Journal of Public Health* 45(5): 1–63.

Paul Meier. 1957. "Safety Testing of Poliomyelitis Vaccine." *Science* 125(3257): 1067–1071.

D. D. Rutstein. 1957. "How Good Is Polio Vaccine?" *Atlantic Monthly* 199:48–51.

Glossary of Statistical Terms

Addition rule – When the probabilities of events A and B are known, the probability that either one *or* the other occurs is the sum of their probabilities minus the probability of their joint occurrence. P(A or B) = P(A) + P(B) – P(A and B).

Alpha (α) – In statistics we have three different uses of the term α: (1) the type I error rate (hypothesis testing); (2) the fixed effect of factor 1 (ANOVA model); and (3) the parametric value for the Y intercept (regression model).

Analysis of Variance – ANOVA is a powerful procedure that allows testing the equality of multiple means. ANOVA can be extended to simultaneously test the effects of two or more factors.

Assumptions – Conditions (rules) that must be true for statistical tests to work correctly. For example, many tests require independence, normality, and equal variances. If these assumptions are not met, we can not be sure that the type I error rate is what we believe it to be.

Beta (β) – As for alpha above, we have three different meanings for this term: 1) the type II error rate; 2) the fixed effect of factor 2 (two-way ANOVA model); and 3) parametric value for the slope (regression model).

Binomial distribution – For trials having two possible outcomes (success or failure) and repeated n times, the set of probabilities for each possible number of successes $x\{0, 1, 2, ..., n\}$. For example, in families of 4 children, the probabilities of 0, 1, 2, 3, 4 girls. The sum of all these probabilities equals 1.

Census – a collection of the entire statistical population (e.g., the heights of every individual in a classroom).

Coefficient of Variation – Standard deviation divided by the mean and multiplied by 100%.

Complement – If we know the probability of an event occurring (p), then the probability that the event does not occur (q) is equal to $1 - p$. This is also called the **subtraction rule**.

Conditional probability – The probability of the first event changes the probability of the second event.

Confidence interval – The range that brackets the true value of a parameter with some known probability. For example, the 95% confidence interval for the

mean (μ) should, in theory, bracket μ in 95% of samples that are collected.

Control group – A group that has been treated in exactly the same way as the experimental group in all ways except for the specific treatment whose effect is being tested (e.g., the placebo control in the polio vaccine trial; see Meier 1989).

Correlation – A statistical procedure that determines how strong is the association between two continuous variables. Compare with *regression* below.

Critical value – A number from a specific probability distribution to which the test statistic is compared. The critical value is based upon the number of degrees of freedom and the type 1 error rate, α. For most procedures, if the test statistic is greater than the critical value, then the null hypothesis is rejected.

Cumulative probability – The sum of all the probabilities less than or equal to some value x. For example, in families of 4 children, the probability of having a family of less than or equal to 3 girls would be $P(x \leq 3) = P(0) + P(1) + P(2) + P(3)$. Note that, because all the probabilities must sum to 1, $P(x \leq 3) = 1 - P(4)$.

Estimation – Using samples and calculated statistics to describe parameters, with some known certainty (e.g., see confidence interval above).

Expected number – the probability multiplied by the sample size. E(X) = n*P(X)

Expected value – In ANOVA, the expected value for each group is simply each group mean. In regression, the expected value of Y_i for any value of X_i is determined from the formula $Y = a + bX$.

Extrapolation – Estimating Y from X for values of X beyond the range of X which was used to establish the regression.

Factor – An independent variable which we manipulate in an experiment. By modifying this factor, we can test whether different *levels* lead to the same mean response by some dependent (response) variable.

Fixed effects – a factor in an ANOVA whose different values were chosen by the experimenter (e.g., different types of fertilizer used in an experiment testing plant growth).

Frequency distribution – a table or graph that illustrates the number of occurrences (frequency) for different values of a variable. Usually the variable is a measurement. For continuous measurements, this graph is called a *histogram*; for discrete (integer) measurements, this is a *bar graph*.

Histogram – a graph of a frequency distribution for continuous data.

Homogeneity of variance – An important assumption of both ANOVA and regression, that the within-groups variances are the same for all groups. Also called "homoscedasticity."

Hypothesis – An idea that can be tested by observation. The *working hypothesis* is what we believe to be correct, and allows us to erect *statistical hypotheses* (null and alternative hypotheses), that can then be directly tested by comparing a test statistic with a probability distribution.

Independent events – Two events, A and B, are independent when the occurrence of event A has no influence on the probability of event B occurring and when event B has no influence on the probability of event A occurring.

Inference – Drawing a conclusion about a statistical population based on a sample from that population. Inferential statistics include estimation and hypothesis testing.

Interaction – In a two-way ANOVA, an interaction occurs when the effect of one factor depends upon the level of the second factor (and vice versa). When an interaction occurs, the main effects are not strictly additive.

Interquartile range – the range of the middle 50% of a set of numbers, ordered from smallest to largest (i.e., the difference between the first and third quartiles).

Levels – Different treatment groups. For example, in a one-way ANOVA, if we have three different fertilizer treatments, there are three levels of one *factor* (fertilizer).

Mean – the arithmetic average.

Median – the middle value in an ordered array.

Mode – the most common value; this would represent the peak of a frequency distribution.

Multiplication rule – When events A and B are independent, the probability of both events occurring is equal to the probability of A times the probability of B. The general rule is P(A and B) = P(A)*P(B|A). ["P(B|A)" means the probability of B, given that A has occurred.]

Mutually exclusive – Two events are mutually exclusive when the occurrence of one event prohibits the occurrence of the other event.

Null hypothesis – The statistical hypothesis which generates a probability distribution, with which a test statistic can be compared. Commonly, the null hypothesis represents the conservative position (e.g., X had no effect on Y). *Note of caution:* when we fail to reject the null hypothesis, we have *not* proven that the null hypothesis is true; we can only say the data are consistent with this hypothesis.

Outlier – a value in a sample of measurements which differs greatly from all the rest. An outlier can greatly change the values of the mean and standard deviation.

Parameter – the true value for some characteristic of a statistical population. (e.g., the mean, μ, or variance, σ^2). A parameter is ordinarily *estimated* by calculating a statistic from a *sample* (e.g., the sample mean and sample variance).

Poisson distribution – For events which are rare but have the potential of occurring a large number of times, when the events are randomly distributed in space or time, their probabilities follow a Poisson distribution.

Population – see statistical population below.

Power of the test – The probability of rejecting a null hypothesis when it is indeed false. Equal to $1 - \beta$. Power of the test depends on α, the sample size (n), and the effect size (the difference between the hypothesized mean and the true mean).

Precision – the closeness of repeated measures to one another. The precision of a continuous measurement depends on the sensitivity of the measuring instrument (e.g., a truck scale vs. Mettler balance). This precision is implied by the number of significant figures which are recorded.

Probability – the risk or chance that a particular event ("*success,*" S) will occur in an uncertain world. If an experiment is repeated an infinite number of times, the probability of S is the proportion of times that S occurs. Probability may be determined in two general ways: (1) the classical approach, where we have a prior expectation, based on knowing the "rules of the game"; and (2) the empirical approach, where we know the relative frequency of the event from having done extensive surveys.

Probability distribution – a listing of all possible values of a random variable, together with the probability of each value (e.g., Binomial distribution).

p-value – The probability of the observed outcome (or more extreme), if the null hypothesis were true. When the p-value is small, we will reject the null hypothesis.

Quartiles – measures of location which divide an ordered set of numbers into quarters. The first quartile (Q1) is the value which divides the lower 25% from the rest; the third quartile is the value which divides the upper 25% from the rest; and the second quartile is identical with the median (divides the lower from upper 50%).

Random effects – a factor in an ANOVA whose different values represent a random sample from a natural distribution of populations (e.g., a random selection of fish populations from a large group of lakes).

Random sample – each sample of the same size has an equal probability of being drawn.

Regression – A statistical procedure that models the dependence of one (dependent) variable on changes of a second (independent) variable. Both the dependent (response) and independent (predictor) variables are continuous measurements.

Residual – the unexplained error; the difference between a measurement and its expected value.

Response variable – The dependent variable in any experiment. We ask the question "Does variable Y depend upon our manipulation of some factor X (independent predictor variable)?"

Sample – a subset of the statistical population, generally collected at random. Ordinarily, we make inferences about the population, based upon properties of a sample.

Science – investigation of rational concepts that can be tested by observation and experiment. Science also refers to a body of facts, e.g., biology. However, the key characteristic of all sciences is the application of the scientific method.

Scientific method – the cycle of observation, generalization ("invention" or induction), prediction (deduction), experimentation, and statistical validation, which serves as the basis of all sciences.

Standard deviation – the "average deviation" from the mean. A measure of variability, which equals the square root of the *variance*.

Standard error of the mean – If a large number of samples, each of size n, were collected and the sample mean calculated for each sample, the standard error (SE) is the standard deviation of these means. In ordinary usage, the SE is calculated as the sample standard deviation (s) divided by the square root of the sample size (n). The SE represents the precision of the estimate of the mean; precision improves (variation becomes less) as the sample size increases.

Statistics – the word "statistics" has two meanings, both correct. First, statistics refers to descriptive properties from samples (e.g., batting average of a baseball player; variance in blood sugar levels in Hispanic men). Second, statistics refers to inferential procedures, used to make decisions about hypotheses (e.g., *t* test, ANOVA).

Statistical population – all the possible observations of some property of interest (e.g., resting heart rates of every 20-year-old female on Earth). If we want to make a generalization about the property, we ordinarily collect a random sample from the population and calculate statistics from that sample.

Test statistic – The general name for such quantities as t_s or F_s, which are calculated from samples, and then compared to a known probability distribution (e.g., t or F) and used for statistical inference.

***t* test** – A statistical procedure for comparing the equality of a mean to a hypothesized value or two means to each other.

Theory – a hypothesis which has been repeatedly tested and supported in every case. A good theory helps to explain many different kinds of observations (e.g., the double helix model of DNA).

Treatment – In an experiment, the variable we are manipulating. Same as *factor* or *independent variable*. Different groups of the factor are called *levels*.

Two-way ANOVA – A factorial design where two independent factors are manipulated simultaneously to examine their effects upon a response variable. Using this design we can examine three different hypothesis tests, the effects of each factor alone and their interaction.

Type 1 error – Rejecting a null hypothesis when it is indeed true.

Type 2 error – "Accepting" (failing to reject) a null hypothesis when it is indeed false.

Variable – a property in which individuals in a sample can differ (e.g., white blood cell density). Statisticians distinguish three main groups of variables: attributes (categories), ordinal (ranks), and measurements.

Variance – the sum of squared deviations divided by degrees of freedom. Equal to the standard deviation squared.

Index